Bernadette G. Callery, PhD
Editor

Collaborative Access to Virtual Museum Collection Information: Seeing Through the Walls

Collaborative Access to Virtual Museum Collection Information: Seeing Through the Walls has been co-published simultaneously as *Journal of Internet Cataloging*, Volume 7, Number 1 2004.

Pre-publication REVIEWS, COMMENTARIES, EVALUATIONS . . .

"ADDS TO OUR UNDERSTANDING OF HOW TO MAKE EFFECTIVE USE OF THE WEB. . . . Assists us to comprehend the differences between online and other kinds of users. . . . This book will stimulate discussion in museums, libraries, and archives about how they can most effectively utilize the Web and avoid its pitfalls."

Richard J. Cox, PhD
Professor of Archival Studies
Department of Library and Information Science
School of Information Sciences
University of Pittsburgh

More pre-publication
REVIEWS, COMMENTARIES, EVALUATIONS . . .

"FILLS A SIGNIFICANT GAP IN THE LITERATURE. This work provides practical advice from experienced professionals on the best ways to avoid hazards and trouble. Complete with examples and frank descriptions, this book will assist many individuals faced with difficult choices as digitization efforts move forward with accelerating speed. As those storage devices fill with text and images, the reports in this book will help many a librarian, curator, and technologist to make educated decisions about their growing digital collections."

Edward J. Valauskas
Follett Chair
Graduate School of Library and Information Science
Dominican University

The Haworth Information Press®
An Imprint of The Haworth Press, Inc.

Collaborative Access to Virtual Museum Collection Information: Seeing Through the Walls

Collaborative Access to Virtual Museum Collection Information: Seeing Through the Walls has been co-published simultaneously as *Journal of Internet Cataloging*, Volume 7, Number 1 2004.

Monographic Separates from the *Journal of Internet Cataloging*™

For additional information on these and other Haworth Press titles, including descriptions, tables of contents, reviews, and prices, use the QuickSearch catalog at http://www.HaworthPress.com.

Collaborative Access to Virtual Museum Collection Information: Seeing Through the Walls, edited by Bernadette G. Callery, PhD (Vol. 7, No. 1, 2004). *A comprehensive discussion about the strategies used by museum archivists, curators, and librarians to provide collaborative online access to their virtual collections.*

High-Level Subject Access Tools and Techniques in Internet Cataloging, edited by Judith R. Ahronheim (Vol. 5, No. 4, 2002). *Examines a state-of-the-art cross-section of research data and practical strategies for developing a high-level browse interface.*

Encoded Archival Description on the Internet, edited by Daniel V. Pitti, CPhil, MLIS, and Wendy M. Duff, PhD (Vol. 4, No. 3/4, 2001). *"A broad overview of EAD: how it came to be, what it does, where it fits with other descriptive standards, and what its future might hold. Thought-provoking articles explore EAD's potential impact on reference service, archival information systems, and museum descriptive practices–areas that have not yet been fully explored or exploited by the archival community." (Kris Kiesling, MILS, Head, Department of Manuscripts and Archives, Harry Ranson Humanities Research Center, University of Texas at Austin)*

CORC: New Tools and Possibilities for Cooperative Electronic Resource Description, edited by Karen Calhoun, MS, MBA, and John J. Riemer, MLS (Vol. 4, No. 1/2, 2001). *Examines the nuts-and-bolts practical matters of making a cataloging system work in the Internet environment, where information objects are electronic, transient, and numerous.*

Metadata and Organizing Educational Resources on the Internet, edited by Jane Greenberg, PhD (Vol. 3, No. 1/2/3, 2000). *"A timely and essential reference. . . . A compilation of important issues and views . . . provides the reader with a balanced and practical presentation of empirical case studies and theoretical elaboration." (John Mason, Co-Chair, Dublin Core Education Working Group, and Technical Director, Education, Au LTD [Education Network Australia])*

Internet Searching and Indexing: The Subject Approach, edited by Alan R. Thomas, MA, and James R. Shearer, MA (Vol. 2, No. 3/4, 2000). *This handy guide examines the tools and procedures available now and for the future that will help librarians, students, and patrons search the Internet more systematically, and also discusses how Internet pages can be modified to facilitate easier and efficient searches.*

Collaborative Access to Virtual Museum Collection Information: Seeing Through the Walls

Bernadette G. Callery
Editor

Collaborative Access to Virtual Museum Collection Information: Seeing Through the Walls has been co-published simultaneously as *Journal of Internet Cataloging*, Volume 7, Number 1 2004.

The Haworth Information Press®
An Imprint of The Haworth Press, Inc.

New York • London • Victoria (AU)
www.HaworthPress.com

Published by

The Haworth Information Press®, 10 Alice Street, Binghamton, NY 13904-1580 USA

The Haworth Information Press® is an imprint of The Haworth Press, Inc., 10 Alice Street, Binghamton, NY 13904-1580 USA.

Collaborative Access to Virtual Museum Collection Information: Seeing Through the Walls has been co-published simultaneously as *Journal of Internet Cataloging*™, Volume 7, Number 1 2004.

Cover design by Jennifer M. Gaska.

Library of Congress Cataloging-in-Publication Data

Collaborative access to virtual museum collection information : seeing through the walls / Bernadette G. Callery, editor.

p. cm.

"Co-published simultaneously as Journal of Internet Cataloging, Volume 7, Number 1 2004."

Includes bibliographical references and index.

ISBN-13: 978-0-7890-2933-1 (hc. : alk. paper)

ISBN-10: 0-7890-2933-2 (hc.: alk. paper)

ISBN-13: 978-0-7890-2934-8 (pbk. : alk. paper)

ISBN-10: 0-7890-2934-0 (pbk. : alk. paper)

1. Virtual museums. 2. Digital communications. 3. Museums–Collection management. 4. Electronic information resources–Access control. I. Callery, Bernadette G.

AM7 .C576 2004
069'.5–dc22

2005001501

Indexing, Abstracting & Website/Internet Coverage

This section provides you with a list of major indexing & abstracting services and other tools for bibliographic access. That is to say, each service began covering this periodical during the year noted in the right column. Most Websites which are listed below have indicated that they will either post, disseminate, compile, archive, cite or alert their own Website users with research-based content from this work. (This list is as current as the copyright date of this publication.)

<u>Abstracting, Website/Indexing Coverage</u> . . . . <u>Year When Coverage Began</u>

- ***AGRICOLA Database (AGRICultural OnLine Access) A Bibliographic database of citations to the agricultural literature created by the National Agricultural Library and its cooperators <http://www.natl.usda.gov/ag98>*** **1997**
- ***AGRIS <http://www.fao.org/agris/>*** **1997**
- ***Applied Social Sciences Index & Abstracts (ASSIA) (Online: ASSI via Data-Star) (CDRom: ASSIA Plus) <http://www.csa.com>*** **1997**
- ***CareData: the database supporting social care management and practice <http://www.elsc.org.uk/caredata/caredata.htm>*** . . . **2001**
- ***Combined Health Information Database (CHID)*** **1997**
- ***Computer and Information Systems Abstracts <http://www.csa.com>*** **2004**
- ***Computer Science Index (CSI) (formerly Compter Literature Index) (EBSCO) <http://www.epnet.com>*** **1997**
- ***Current Cites [Digital Libraries] [Electronic Publishing] [Multimedia & Hypermedia] [Networks & Networking] [General] <http://sunsite.berkeley.edu/CurrentCites/>*** **1998**

(continued)

- *EBSCOhost Electronic Journals Service (EJS) <http://ejournals.ebsco.com>* . . . **2001**
- *FRANCIS. INIST/CNRS <http://www.inist.fr>* . . . **1999**
- *Google <http://www.google.com>* . . . **2004**
- *Google Scholar <http://scholar.google.com>* . . . **2004**
- *Haworth Document Delivery Center* . . . **1997**
- *IBZ International Bibliography of Periodical Literature <http://www.saur.de>* . . . **2002**
- *Index to Periodical Articles Related to Law <http://www.law.utexas.edu>* . . . **1997**
- *Information Science & Technology Abstracts: indexes journal articles from more than 450 publications as well as books, research reports, and conference proceedings; EBSCO Publishing <http://www.epnet.com>* . . . **1997**
- *Informed Librarian, The. For more information visit us at: <http://www.informedlibrarian.com>* . . . **2003**
- *INSPEC is the leading English-language bibliographic information service providing access to the world's scientific & technical literature in physics, electrical engineering, electronics, communications, control engineering, computers & computing, and information technology <http://www.iee.org.uk/publish/>.* . . . **1997**
- *Internationale Bibliographie der geistes- und sozialwissenschaftlichen Zeitschriftenliteratur . . . See IBZ <http://www.saur.de>* . . . **2002**
- *Internet & Personal Computing Abstracts (IPCA) (formerly Microcomputer Abstracts) provides access to concise and comprehensive information on the latest PC products & developments, covering over 120 of the most important publications; EBSCO Publishing <http://www.epnet.com/public/internet&personal.asp>* . . . **1999**
- *Konyvtari Figyelo (Library Review)* . . . **1997**
- *Library & Information Science Abstracts (LISA) <http://www.csa.com>* . . . **1997**
- *Library and Information Science Annual (LISCA) <http://www.lu.com>* . . . **1999**
- *Management & Marketing Abstracts <http://www.pira.co.uk/>* . . . **2000**

(continued)

- *MLA International Bibliography provides a classified listing & subject index for books & articles published on modern languages, literatures, folklore & linguistics. Available in print and in several electronic versions. Indexes over 50,000 publications <http://www.mla.org>* **1997**
- *OCLC Public Affairs Information Service <http://www.pais.org>* **1999**
- *PASCAL, c/o Institut de l'Information Scientifique et Technique. Cross-disciplinary electronic database covering the fields of science, technology & medicine. Also available on CD-ROM, and can generate customized retrospective searches <http://www.inist.fr>* **1998**
- *Referativnyi Zhurnal (Abstracts Journal of the All-Russian Institute of Scientific and Technical Information– in Russian) <http://www.viniti.ru>* **1997**
- *Social Work Abstracts <http://www.silverplatter.com/catalog/swab.htm>* **1997**
- *Sociological Abstracts (SA) <http://www.csa.com>* **1997**
- *SwetsWise <http://www.swets.com>* **2001**
- *World Publishing Monitor* **2000**
- *zetoc <http://zetoc.mimas.ac.uk/>* **2004**

Special Bibliographic Notes related to special journal issues (separates) and indexing/abstracting:

- indexing/abstracting services in this list will also cover material in any "separate" that is co-published simultaneously with Haworth's special thematic journal issue or DocuSerial. Indexing/abstracting usually covers material at the article/chapter level.
- monographic co-editions are intended for either non-subscribers or libraries which intend to purchase a second copy for their circulating collections.
- monographic co-editions are reported to all jobbers/wholesalers/approval plans. The source journal is listed as the "series" to assist the prevention of duplicate purchasing in the same manner utilized for books-in-series.
- to facilitate user/access services all indexing/abstracting services are encouraged to utilize the co-indexing entry note indicated at the bottom of the first page of each article/chapter/contribution.
- this is intended to assist a library user of any reference tool (whether print, electronic, online, or CD-ROM) to locate the monographic version if the library has purchased this version but not a subscription to the source journal.
- individual articles/chapters in any Haworth publication are also available through the Haworth Document Delivery Service (HDDS).

∞ ALL HAWORTH INFORMATION PRESS
BOOKS AND JOURNALS ARE PRINTED
ON CERTIFIED ACID-FREE PAPER

Collaborative Access to Virtual Museum Collection Information: Seeing Through the Walls

CONTENTS

ABOUT THE EDITOR

Bernadette G. Callery, PhD, is Museum Librarian and Head, Library and Archives, at the Carnegie Museum of Natural History. Her research interests include the development of museum recordkeeping systems, particularly as they move from paper-based systems to electronic ones. For the last five years, she has served as a visiting lecturer at the University of Pittsburgh's School of Information Sciences, where she teaches the Museum Archives course. Active in the Society of American Archivists' Museum Section, Dr. Callery has lectured and published on various aspects of museum archives, including contributing to *Museum Archives: An Introduction*, edited by Deborah Wythe and published by the Society of American Archivists in 2004. Based on experience gained during her previous positions at the Hunt Institute for Botanical Documentation and the New York Botanical Garden, she continues to lecture on the history of botanical illustration.

Foreword

The essays in this volume are an important contribution to our understanding of the possibilities–and the problems–of working collaboratively to develop a new approach to enhancing knowledge about and access to museum collections. Accordingly, these essays are directly relevant to the strategic approach of the Institute of Museum and Library Services (IMLS), for two reasons.

First, at IMLS, one of our major efforts is to foster collaboration. We believe that effective collaboration is *the* strategy of the 21st century. It is aligned with how we are thinking about our communities as "holistic" environments, as social ecosystems in which we are part of an integrated whole. The kind of collaboration we try to foster is simply a mature and reflective recognition of intersecting nodes of interest, activity, and mission.

Second, at IMLS we are dedicated to the purpose of creating and sustaining a nation of learners. We perceive that museums are primarily social agencies intended to provide resources and services that stimulate learning. Museums complement and support the agencies of formal education: schools, colleges, and universities. But equally important, they serve as rich resources for the independent, free-choice learner. They provide unique opportunities and infrastructure to support learners of all ages, in every circumstance. To maximize their ability to function in this way, it is imperative for museum professionals to develop new approaches that maximize community awareness of, and access to, the rich resources in the collections of all museums. Beyond that, such systems must also interoperate to connect museum resources to comparable and correlated resources in other cultural heritage institutions, such as libraries and archives.

[Haworth co-indexing entry note]: "Foreword." Martin, Robert S. Co-published simultaneously in *Journal of Internet Cataloging* (The Haworth Information Press, an imprint of The Haworth Press, Inc.) Vol. 7, No. 1, 2004, pp. xvii-xix; and: *Collaborative Access to Virtual Museum Collection Information: Seeing Through the Walls* (ed: Bernadette G. Callery) The Haworth Information Press, an imprint of The Haworth Press, Inc., 2004, pp. xiii-xv.

We often hear it said today that we are living in an information age. But in a world drowning in information, we are hungry for knowledge. That is why today, in the 21st century, we must be more than an information society. We must become a *learning* society.

A learning society requires that we do more than develop the hardware, software, telecommunications networks, and other services and systems that supply and organize content. It requires additional structure and context to enable learners around the globe to put knowledge to good use.

In the 21st century, environmental conditions mandate that the ability to learn continuously throughout the lifetime is essential. Accelerating change has become a way of life. To navigate such change, minimize risk, and participate effectively in civic affairs, all Americans need access to learning throughout their lifetimes. The importance of continuous learning, free-choice learning, life-long learning, for economic vitality and for personal fulfillment alike, is beyond question.

To support such learning, we also need to envision a seamless learning infrastructure, one that stimulates and supports learning for learners of all ages, in all circumstances. If we are truly to empower individuals to fulfillment, to enable them to maximize their human potential, to become contributing members of the knowledge economy, and to participate effectively in civic affairs, then we must build a fabric of social agencies that facilitates continuous lifelong learning. Such a fabric must integrate all of our current social agencies that create, manage, and provide access to learning resources, including schools, libraries, museums, archives, broadcasters, and a host of other organizations and agencies.

Digital technology has provided the means to do just that. In the online environment, the boundaries that separate one museum from another disappear. Resources that our audiences discover online can be re-ordered to meet the needs of the user. They can be deconstructed, and reassembled in response to specific user needs. People who find such resources online often do not know, and frequently do not care, what kind of institution houses the original material. They simply want access to these rich resources to address their own perceived needs.

But locating authentic and useful material online, among the vast amount of junk that populates the Internet, is not an easy task. Learners want material that is authentic, that is reliable, that has validity and relevance to their own individual needs. They can find such things in online museum collections. But we must make it easier for them to locate and retrieve such resources. We need enhanced awareness and discovery tools, leading independent learners to the materials that they value. Such systems will not only empower users to find individual discrete resources that are relevant to their needs, but also enhance the ability to search across domains to co-locate relevant resources.

The core of any such awareness and discovery tools will be the development of collaborative cataloging systems, built on commonly agreed descriptive standards and protocols, that will enable cross-platform searching and retrieval. While this kind of collaboration can be a challenge to our traditional and normal modes of operation, it is imperative that we embrace the challenge in order to build a seamless infrastructure for learning that will be valued by the communities that we seek to serve.

Museums–as well as libraries, archives, and other cultural agencies–are indispensable elements in this web of learning. Given our recent history of collaboration and our tradition of service, we can lead the way in demonstrating the potential and developing the reality.

The essays in this volume are an important contribution to our understanding of the problems and possibilities of building the systems and services that our audiences demand. They highlight the dramatic potential inherent in our collections. They also reveal some of the pragmatic limitations in the landscape that will require significant effort and hard work to overcome. Most exciting, perhaps, is the promise for developing systems that support interactive engagement with the audience for our collections, inviting the contributions of the museum-goer in developing new perspectives on the objects and resources that we preserve and provide access to.

This volume breaks new ground. It offers challenges, illuminates potential, and invites innovation. The authors and editors are to be congratulated for what they have contributed.

Robert S. Martin
Director
Institute of Museum and Library Services
Washington, DC
E-mail: rmartin@imls.gov

Introduction

Bernadette G. Callery

Museums are about interpretation–that interaction that takes place between people, objects, and information in the particular intellectual and social environment of the museum. The interpretation of objects in museums, whether paintings, butterflies, or railroad engines, includes the largely invisible context of the associated metadata extracted from such collection records as donor's correspondence files, collector's field notes, and conservation records dealing with painted metal surfaces. Museums also include the visitor, who interacts with the objects and their information. Assuming that the "virtual museum" is not an oxymoron, one of the challenges of the virtual museum is to translate this essential aspect of interaction into an online environment, where the online audience may have very different expectations of the museum experience than the audience walking through the doors. Creators of virtual museum projects, including the ones discussed in this collection of papers, have been exploring some of the descriptive standards and technologies already in use by libraries to allow museums to extend that experience of interpretation to the offsite audience. However, the use of separate, but not necessarily equal, conventions of record structure, data elements and descriptive terminology between the museum and library community can result in some confusing experiences and lost opportunities, particularly in the collaborative online environment.

[Haworth co-indexing entry note]: "Introduction." Callery, Bernadette G. Co-published simultaneously in *Journal of Internet Cataloging* (The Haworth Information Press, an imprint of The Haworth Press, Inc.) Vol. 7, No. 1, 2004, pp. 1-8; and: *Collaborative Access to Virtual Museum Collection Information: Seeing Through the Walls* (ed: Bernadette G. Callery) The Haworth Information Press, an imprint of The Haworth Press, Inc., 2004, pp. 1-8. Single or multiple copies of this article are available for a fee from The Haworth Document Delivery Service [1-800-HAWORTH, 9:00 a.m. - 5:00 p.m. (EST). E-mail address: docdelivery@haworthpress.com].

http://www.haworthpress.com/web/JIC

Digital Object Identifier: 10.1300/J141v07n01_01

ISSUES OF COLLABORATIVE ACCESS

These papers present a sample of the options for providing access to museum collection information, and range from online scrapbooks combining selected images and interpretive text in the Ohio Memory Online Scrapbook and the King County Snapshots to technologies used in the Experience Music Project for recreating the performance experience of popular American music (think Jimi Hendrix, not Stephen Foster). Some of the gaps between vision and reality have been identified–and bridged–in these papers, largely case studies of collaborative construction of virtual museum collection catalogs. The first challenge is the disparity of the vocabularies used by the individual museums to describe their collections. Not only do museums have collections of unique materials, but they also have different traditions of nomenclature, different hierarchical classifications of information, and different conventions for reporting such essential descriptors as date and location. Many online museum catalogs were designed initially to serve the needs of curatorial and collection management staff and reflect the highly specialized terminology of those scientific and artistic fields. Individual museums also have their own information cultures, which determine how individual departments of the museum interact with the non-scholarly public. The nature of this interaction determines how museums will approach the second challenge, which is the tradition of mediated access to museum information. The most common example of mediated access is the museum exhibit, in which the objects, their arrangement, and their immediate meaning, are completely selected, organized, and interpreted by the museum staff. Onsite visitors wishing further information are directed to the museum's curatorial staff who will determine, on a case by case basis, the user's need to know, then select and interpret records, objects, and other information for their use. Offsite visitors to the museum's Web site are generally expected to content themselves with the Web-searchable collection information but may have the option of sending in their specific questions to a generic e-mail address.

The virtual museums discussed here, particularly the multi-museum collaborative projects, have addressed the first challenge by adhering to a project-based set of standard terminologies, largely through the acceptance and use of a centralized authority. They have approached the second in the spirit of compromise, by allowing each contributing museum to select the specific type of information included in each object's record. The collaborative projects are primarily intended to serve as a means of remote information discovery, and are particularly valuable when objects once associated with a single person or event would not come together in the reality of a specific museum's holdings. In a virtual museum they do. Collaborative catalogs open the collections of all the participating museums to further interpretation as a group. One of the distinguishing aspects of these projects is the opportunity for the creators or performers to speak directly to the user of the online museum as in the recorded interviews and sound bites in the Experience Music Project. There are also opportunities for descendants of the creators of ethnographic objects to speak directly to the user through their own additions to the existing museum catalog record as in the virtual tours developed by Native children working with tribal collections in the National Museum of the American Indian.

A common initial expectation in the creation of online museum catalogs is that existing cataloging conventions and practices can simply be recast in the electronic environment. As noted in Chen's study of six museums and their digitized and online collections of Chinese material, the procedures and tools used in the past, particularly those designed for the management of slide collections, do not translate well to the online environment. His findings suggested that while the local traditions and expectations of mediated access to slide library collections "satisfy user's needs and minimize the insufficiency of indexing tools and standards, the current tools and standards are not transferable to the digital image collections." Collaborative image management, including the ability to search simultaneously across multiple sites using a single search key continues to be a problem due to inconsistencies in access terminology.

One of the fears of participation in collaborative catalogs is loss of local control over the distribution of sensitive information. What exactly constitutes sensitive information varies across disciplines and individual museums; however, donor, purchase, and provenance information is often considered sensitive within the context of art museums, as location information of artifacts and specimens is for natural history museums. Callery's paper examines the various techniques for identifying sensitive information and alerting prospective online users to its partial or complete suppression from the public catalog records. Several collaborative projects have dealt with the problem by allowing institutions contributing information to a centralized resource to make their own decisions about what information–or what level of detailed information–to contribute. Typically, the user searches the contents of the collaborative catalog to take advantage of its many datapoints for access, but is linked back to the individual holding museum for detailed provenance and other descriptive information and contacts for rights management. The Nazi-Era Provenance Internet Portal, designed to improve access to provenance information on art that changed hands in Continental Europe during the Nazi era (1933-1945), is an example of such a collaborative tool intended to serve the needs of both research and repatriation.[1]

SCRAPBOOK FEATURES

The motif of the scrapbook as used in a number of these collaborative museum projects can be seen as the user's opportunity to create his own interpretation of a museum's collections. The opportunity to establish one's own order on a collection by selecting and arranging objects of personal interest is a popular option in Internet-based museum catalogs. The Ohio Memory Online Scrapbook, King County Snapshots, and the Experience Music Project each offer such a feature. In addition, the scrapbook theme was chosen because of its connotations of a collection of material, often in disparate formats and accumulated over time that chronicle an individual, group, or event. The "my scrapbook" feature in the Ohio Memory Online Scrapbook is typical in that it allows users the ability to make their own collection from the items available in the on-line catalog and annotate them, essentially creating their own museum exhibition catalog. Beyond the obvious opportunity to encourage individual user engagement with the collections, several of these collaborative projects used the scrapbook option to create pre-built "scrapbooks" in order to focus attention on such popular themes as "First Flight" in the Ohio Memory Online Scrapbook.

The King County Snapshots project also used this approach when offering the pre-built searches to pull together material under such topics as "Parades & Festivals" and "Military & War: Armories." In addition to highlighting topics of particular interest in the combined collections in the Seattle area, Janney also noted that these complex queries "could probably not be successfully performed by users unfamiliar with the cataloging used in specific images." These potted searches demonstrate the strength of the integrated system, showing the value and utility of these searches across the entire database.

The Experience Music Project also allows the user to personalize his experience as a part of the onsite visit with the "My Bookmarks" feature, an electronic artifact bookmarking system for visitors using the electronic Museum Exhibit Guide (MEG). When browsing in the galleries, users make note of objects of particular interest with the MEG, then use these customized lists to query the museum's Digital Collection for further details. This system is available in both the onsite and Web versions of the Digital Collection and serves to extend the experience and exploration of the collection even after the visitors had left the building.

Perhaps the ultimate in visitor interaction are the virtual tours of the collections of the National Museum of the American Indian designed by Native children who were given direct access to the Museum's collections. The children made three-dimensional movies of various objects de-installed from existing museum exhibits and added their own captions and catalog descriptions, thus creating virtual exhibitions that are available to the public on the museum's Web site. Perhaps all of these efforts at the personalization of the online museum catalogs are the electronic extension of the human urge to accumulate and organize objects to give them additional meaning.

COLLABORATIVE STANDARDS

Successful collaborative virtual museum projects often begin with existing data but then review it for consistency or adaptability to standards. One approach, taken by the Ohio Memory Online Scrapbook project, was to invite the state's institutions to contribute images and associated catalog records to a central point for review and input. Although Nelson and Gemmill note that "more than 320 cultural heritage repositories in Ohio contributed records to a common electronic catalog without even knowing it," that is not strictly true as those contributing information realized that they were participating in a common catalog, even if they would not create the final record. The goal was integration through standardization, resulting in the ability to map data across different systems. In this case the central authority was based at the Ohio Historical Society which was responsible for the consistency of description, particularly in the choice of subject terms applied to individual objects. The added value of this centralized leadership was that it addressed imbalances of experience, equipment, and the ability to maintain the new resource of electronic collection information. The goal was access to the great range of visual material and the delivery of a rich image-based result set in response to a search. For this reason, all objects were illustrated and records for handwritten documents included a page image as well as transcriptions. Nelson and Gemmill noted that "Linking to images reduces emphasis placed on descriptive metadata, because the records are no longer merely surrogates for original materials; their value lies more in supplementing and contextualizing collections, providing background not just on the item, but on the person, place, object, and event. The search process is not an interim step on resource discovery but a direct connection to primary source materials and historical insight."

Direct access to images was also the goal of the King County Snapshots project, and this was achieved using CONTENTdm and a collaborative approach to metadata development, headed by specialists at the University of Washington Libraries and the Museum of History and Industry (MOHAI), Seattle. Janney's article on this project discusses the difficulties of achieving and maintaining metadata standards, given that several of the twelve project participants already had well-established thesauri for their specific subject-based collections. Cataloging standards for the project were developed by the project manager and the two metadata specialists who traveled to each of the participating organizations following cataloging training workshops designed as part of the project. Existing and desired access points were discussed between the participants and the metadata specialists and mapped to Dublin Core elements. The University of Washington's already established Metadata Implementation Group offered valuable support in the process of developing data dictionaries for each of the partner organizations. Irreconcilable differences between the authorized subject terms and local vocabularies were accommodated by including the local terminology in the Caption or Notes fields, which mapped into the Dublin Core Description element. Caption fields were also used to hold interpretive data supplied by the collaborating museums similar to that found on an exhibition label. With CONTENTdm, users at each organization can limit their searches to their own institution or search across all participants via the project Web site. Since each collection is created as a separate database, it can have any data fields its creators choose, although there were clear advantages to using those that could be mapped to Dublin Core.

The discussion of the Experience Music Project (EMP) by Maguire, Motson, Wilson, and Wolfe presented unusual challenges of including contextual information on what might appear to be ordinary musical instruments or sheet music but were in fact a guitar smashed onstage by Kurt Cobain and the original music for "Louie Louie." In this catalog, information on provenance and circumstances of use are extremely significant pieces of information. The EMP's mission was to explore creativity as expressed in American popular music, with a specific intent to engage the visitor in that creative process. Public availability of detailed information on the museum's collections, including images and audio and video clips, was a stated intent of the project from the beginning. Unlike many other electronic cataloging projects in museums which are designed to primarily serve the internal users as a collection management tool, this catalog was initially designed to also deliver the exhibition content to the public user. This online equivalent of an exhibition catalog was also to include as much of a virtual experience of the objects and performances as the technology could provide. The resultant product combined elements from more than one standard developed by libraries, archives, and other cultural heritage repositories, all incorporated into the EMP cataloging manual. Individual records, which are created for both the object and its digital surrogate, typically include the History Remarks field, which are extended notes on the context in which the object functioned and are often substantial essays which build on the museum's extensive collection of oral history transcripts of donors and performers. Comparable to the History Remarks field, the name authority files also provide biographical information on various bands, in addition to recording earlier and later forms of the group's name. An extensive in-house thesaurus is also maintained in support of the cataloging project. Jim Fricke, the curator of the collection envisioned that the collection database would be "the informational lifeline of the museum, functioning behind the scenes as an organizational, management, and research tool and on the exhibit floor as a guide through the museum's collection and informational resources." The project is an object lesson in how the interpretation of the collection defines the museum and is a remarkable use of re-purposing of collection information for both onsite and online exhibition purposes.

ONLINE INTERPRETATION

A striking example of user-supplied interpretation of museum collections is discussed in the projects hosted by National Museum of the American Indian (NMAI). In a collaboration between the NMAI and various groups of Native children, their educators, and Indian community members, the children are brought to the Museum's collections and create virtual exhibitions by photographing the objects, then researching and writing captions for the objects chosen for their virtual tour. As Christal, Roy, and Cherian remark, "the students were becoming the 'storytellers for their grandparents,' or 'representatives of the future' for the tribe." These tours are available online to all online visitors through the Museum's Web site and are particularly welcomed by educators looking for source material from the perspective of Native people. The paper includes extensive quotations from the participants in the various workshops which make it clear that this is a powerful opportunity for the users to both experience these items of their cultural heritage directly and leave their reinterpretation for others to experience.

Adding to the growing literature of the repatriation of other cultural heritage items, such as Nazi-looted art, are those works which deal with the responsibilities of museums to respect the culturally-sensitive objects in their collections. This care can extend across many aspects of collection management from cataloging and description to storage and conservation. In Clavir's recent work on the changing attitudes of museum conservators to the care of Native American material, she includes the remarks of a Native museum director on the division of responsibilities between modern conservators and Native peoples who notes, "Your job is to preserve those 'things.' It's our job to preserve the culture that those 'things' have meaning in."[2]

The willingness of the museum to accept its role as a forum in which multiple viewpoints can be expressed rather than the single authoritative "museum voice" can be best seen in the exhibition label one of the NMAI student groups produced for a collection of feathered dance fans. Instead of creating a text, the students were videotaped giving a performance of their village dances using the fans. Christal, Roy, and Cherian note that "Although this kind of use of fairly fragile objects would be prohibited by standard museum practice, it was deemed worth the risk of damage to reunite the items in this matter with the people who possess such a strong cultural connection to them."

All the papers in this collection deal with the ways in which the many voices of the creators, custodians, and consumers of museum objects can be heard and understood by the larger community of online users.

NOTES

1. Nazi-Era Provenance Internet Portal [Accessed 20 April 2004] Available at URL: http://www.nepip.org/.
2. Naomi Clavir, *Preserving What Is Valued: Museums, Conservation and First Nations.* Vancouver, UBC Press, 2002, 212.

Searching for Nirvana: Cataloging and the Digital Collection at the Experience Music Project

Marsha Maguire
David E. Motson
Gwen Wilson
Jen Wolfe

SUMMARY. The Experience Music Project (EMP) in Seattle is an interactive museum dedicated to rock 'n' roll and other forms of popular

---

Marsha Maguire, MLS, MPhil, is Manuscripts & Special Collections Cataloging Librarian, University of Washington Libraries, PO Box 352900, Seattle, WA 98195 (E-mail: mmaguire@u.washington.edu). She was formerly Archives and Access Librarian, Experience Music Project.

David E. Motson, MLS, is Database Maintenance Librarian/Special Materials Cataloger, Columbia University Libraries, 101M Butler Library, 535 West 114th Street, New York, NY 10027 (E-mail: dm820@columbia.edu). He was formerly Chief Cataloger, Experience Music Project.

Gwen Wilson is Digital Lab Administrator, Experience Music Project, 325 5th Avenue North, Seattle, WA 98109 (E-mail: GwenW@emplive.com).

Jen Wolfe is Cataloging Librarian, University of Iowa Libraries, 100 Main Library, Iowa City, IA 52245 (E-mail: jenw@jenw.org). She was formerly Assistant Cataloger/Librarian, Experience Music Project.

The authors would like to thank Stephen Rostkoski, Cataloger/Researcher; Jim Fricke, Curatorial Manager; Jacob McMurray, Curator; and Dawn Haggarty, Manager, Library & Research, Experience Music Project; as well as Melissa Taylor, formerly Collections Information Specialist, EMP; and Linda Miles for their valued assistance in the preparation of this text.

[Haworth co-indexing entry note]: "Searching for Nirvana: Cataloging and the Digital Collection at the Experience Music Project." Maguire, Marsha et al. Co-published simultaneously in *Journal of Internet Cataloging* (The Haworth Information Press, an imprint of The Haworth Press, Inc.) Vol. 7, No. 1, 2004, pp. 9-31; and: *Collaborative Access to Virtual Museum Collection Information: Seeing Through the Walls* (ed: Bernadette G. Callery) The Haworth Information Press, an imprint of The Haworth Press, Inc., 2004, pp. 9-31. Single or multiple copies of this article are available for a fee from The Haworth Document Delivery Service [1-800-HAWORTH, 9:00 a.m. - 5:00 p.m. (EST). E-mail address: docdelivery@haworthpress.com].

http://www.haworthpress.com/web/JIC

Digital Object Identifier: 10.1300/J141v07n01_02

music. This article briefly traces the history of EMP, lists some highlights of the collection, and discusses the development of rules and tools used in cataloging the artifacts and their digital reproductions. Also included is a description of the EMP Digital Collection, a database available to the public through an onsite Digital Lab and on the Web. A user survey intended to gauge the effectiveness of the Digital Collection concludes the article.

KEYWORDS. Virtual museums, Experience Music Project, EMP, rock 'n' roll, music museums, cataloging, Internet, standards, cataloging of artifacts, interactive museum exhibits

With the best of American rock 'n' roll as its inspiration, the Experience Music Project set out in 1992 to become a force for generating a very different but very accessible sort of museum experience, one that would not only present but stimulate creative expression. The mission of EMP since its establishment by Microsoft co-founder Paul G. Allen has been to explore creativity as expressed in American popular music, especially rock 'n' roll, through a carefully developed collection of artifacts, unique architecture by Frank O. Gehry, state-of-the-art technology, and compelling interactive presentations. As stated on the museum's Web site (www.emplive.com), EMP "provides dynamic, multifaceted, ever-changing experiences through new and exciting explorations of American popular music, which both entertain and engage visitors in the creative process."

This mission of personally and dynamically involving visitors in the creative energy of music informed every activity during the planning phase of the museum, from exhibit and program building to collection development to the selection of the site and the design of the facility. Even the cataloging database was intended to be something special: a repository of information, digital images of the artifacts, and digital audio and video that would be available on workstations in the museum itself and online via the EMP Web site. From the beginning, the cataloging staff determined that the digital presentation of the artifacts and music comprising EMP's collections, as well as the cataloging information that described and provided access to those artifacts, would be based on standard practices. With reliable standards to inform the structure and content of the catalog, and cutting-edge technology to enable its construction and delivery, the collections staff had high hopes of developing a truly useful and innovative resource discovery and presentation tool. We aspired to a kind of professional nirvana right from the start.

This article will briefly describe the programs and collections of the Experience Music Project, the development of rules and tools for cataloging those collections, and the ways in which the collections staff have borrowed from existing standards and techniques to accommodate some of the more complex description and search needs of this unique institution (including some instances of local practice). Finally, we'll discuss the effectiveness of the public EMP Digital Collection, accessible onsite in the museum's Digital Lab and on the Web through EMP's home page, as determined from a user survey conducted in the summer of 2003.

THE MUSEUM AND ITS COLLECTIONS

EMP, first known as the Jimi Hendrix Museum, was originally envisioned as a small gallery of Hendrix memorabilia collected by one of the artist's great admirers, Paul G. Allen, in what had been the Modern Art Pavilion in the 1962 Seattle World's Fair. Because Hendrix grew up in Seattle, Allen also planned to include information on the area's music scene as Hendrix experienced it in his youth. In 1992, as these plans were being formulated, the city became something of a musical mecca due to the popularity of "grunge" rock, which was identified with Seattle and epitomized by Seattle-based bands such as Nirvana and Pearl Jam. Consequently, the scope of the project grew to encompass Seattle grunge and then the rock scene of the Pacific Northwest as a whole. By 1995, when the institution's name was changed to Experience Music Project, Allen, executive director Jody Patton, and the curatorial staff, had expanded the original goals into the current vision of the museum, with emphasis placed on creativity in American pop music generally and rock 'n' roll specifically.

From 1995 to 2000, the museum staff developed the exhibit galleries, educational programs, interactive music and technology equipment, and performance and special events venues. They also built a collection that by opening day totaled over 80,000 artifacts and now tops 100,000 items. The collection is rich and complex, with highlights including:

- *Hendrix*: The world's largest collection of Jimi Hendrix recordings, guitars, posters and photographs, motion picture footage, clothing, and manuscripts, such as handwritten lyrics to many of his songs;
- *Guitars*: A huge collection of guitars, amplifiers, and mixing and recording equipment, some of them previously owned by such artists as Muddy Waters, Bob Dylan, Eric Clapton, Bo Diddley, and Kurt Cobain;
- *Northwest*: Artifacts documenting Seattle's early jazz and R&B scenes, the garage rock era, the development of punk and new wave, and the grunge phenomenon. Items range from the original sheet music for "Louie Louie" to stage costumes worn by Heart, and a guitar smashed onstage by Kurt Cobain;
- *Popular music*: Artifacts illustrating specific styles and genres of popular music, including blues, jazz, soul, gospel, hip hop, reggae, disco, and punk;
- *Archival collections*: The unpublished personal papers of important figures in pop music history, such as former *New York Times* music critic Robert Shelton and West Coast country bassist Fred Maddox, as well as the files and records of organizations like radio stations and record companies;
- *Oral history*: Hundreds of interviews with performers, writers, arrangers, artists, agents, designers, fans, and other people and groups related to the creation, performance, promotion, documentation, interpretation, and enjoyment of popular music; recorded on tape, film, and video; with digitized, transcribed, and edited versions prepared for exhibition purposes;
- *Publications*: A valuable research collection of books, fanzines, newspapers, journals, and vertical subject files;
- *And more*: Fan memorabilia; photographs, posters, and writings by notable pop culture artists; furniture and stage props; and promotional materials of all kinds, like "punk look" hair gel, a Pearl Jam air freshener, and a fan-created portrait of Jimi Hendrix made from pizza boxes.

As EMP's curators built the collections and designed the galleries, the collections staff concentrated on accessioning, tracking, insuring, carefully housing and shelving, and finally cataloging this immense body of varied artifacts. In 1994, EMP hired its first catalogers, including two librarians with MLS degrees and experience in cataloging books, archival and graphic materials, and ethnological museum artifacts. Because the museum had been staffed until that time by a few music scholars and collectors, a part-time programmer, and several administrators who had worked primarily in business settings, the new catalogers were surprised and delighted to discover that a database incorporating ISBD descriptive standards had been built by the curatorial and programming staff. The database, created first in ProCite and then in Microsoft Access, consisted at that time primarily of accession-level records describing Hendrix-related sound recordings, instruments, and posters. But this was just the beginning: curator Jim Fricke, a musician and scholar who had taken the time to educate himself about bibliographic description and indexing, set ambitious goals for the collections management database, known as CIS ("Collections Information System"). Here's how he described his vision of the system:

> The . . . [collections] database will be the informational lifeline of the museum, functioning behind the scenes as an organizational, management, and research tool, and on the exhibit floor as a guide through the museum's collection and informational resources.[1]

CIS was intended, then, to support collection cataloging and management (including acquisitions and rights information); multimedia storage and retrieval (serving as a multimedia archive in the museum and on the Web); and multimedia presentation (feeding kiosk presentations throughout the museum). It was even meant to be a transaction and reporting system, useful for maintaining museum membership planning and reporting. By opening day in June of 2000, the first two functions had been realized with CIS, while the other functions ultimately were handled by separate systems.

Those first two functions were tricky enough, though, with cataloging and artifact tracking alone presenting an enormous challenge. The collection built by EMP curators during the years between the museum's founding in 1992 and its opening at the Seattle Center in 2000 reflected the hybrid nature of the institution. Part museum, part library, part archives–and part something brand new–EMP developed a collection comprised of multiple formats: three-dimensional artifacts, books and serials, sound recordings in various formats, graphic materials, film and video materials in various formats, computer files, and archival collections of personal papers and organizational records. The cataloging of this array of materials needed to incorporate elements considered crucial by museum curators and registrars, such as condition reports and detailed physical descriptions necessary for insurance and exhibition purposes; information that archivists consider essential, like contextual biographical/historical notes and scope-and-content descriptions; and descriptive elements included in library cataloging rules for textual materials as well as non-book formats. Because most analog items–in other words, every item not born digitally–would be presented in digital as well as original analog form, technical information about accompanying digital image, sound, or video files would also need to be supplied.

CONTENT STANDARDS

As the cataloging staff began planning to move the existing MS Access database to the next level of complexity and to develop a written set of cataloging and indexing rules, the decision to base those rules on accepted content standards was easily made. We wanted users to reap the advantages that standards provide: enhanced consistency, interoperability, broad consensus on best practice, and insurance against the obsolescence of the cataloging data. Even though there were no plans to contribute EMP's records to a major utility such as OCLC or RLIN, the records would be viewable on the World Wide Web, and we wanted users to be familiar and comfortable with the basic elements and terminology they would find in EMP's catalog records. Echoing curator Jim Fricke's view that the collections database would serve a variety of functions, the museum administrators stressed that information would be re-used and re-purposed in onsite and Web-based exhibits and educational tools, such as artifact captions in the museum galleries and various topical presentations, "artifact-of-the-day" features, and band genealogies on EMP's Web site. Adhering as closely as possible to accepted standards of description would make it possible to re-purpose information in these and other, as yet unimagined, ways.

But what content standards would be both appropriate and useful for describing and providing access to a collection as varied as EMP's? Moreover, if nearly every artifact in the collection were digitized–and at EMP, every item is considered an artifact because every item is a candidate for exhibition–what descriptive standards could be used to describe both the original artifact and its digital surrogates? There were no easy answers–these are the questions faced by every collection manager who hopes to make resources in multiple formats available both physically–in a library, archives, or museum setting–and digitally on the Internet.

At EMP, we decided to take a sort of mix-and-match approach to cataloging: to borrow the most useful and fitting guidelines for our unique body of materials from a selected group of widely accepted content standards. In describing an unpublished photograph, for example, we might create a record based primarily on *Graphic Materials: Rules for Describing Original Items and Historical Collections* (Washington, DC: Library of Congress, 1982), but include a condition statement based on the museum community's *International Guidelines for Museum Object Information: The CIDOC Information Categories* (Paris: ICOM, 1995).

This approach is not unprecedented, although it was not in common use when we developed EMP's written cataloging guidelines in the mid-1990s. An example of combining elements from more than one standard can be seen in the proliferation in recent years of Dublin Core application profiles. In their 2000 article in the online journal *Ariadne*, Rachel Heery and Manjula Patel define application profiles as "schemas which consist of data elements drawn from one or more namespaces, combined together by implementors, and optimised for a particular local application."[2]

Although it applies more to structural than to content standards, the idea behind the development of an application profile is eminently sensible. Rather than create a strictly homegrown set of descriptive elements and values–thereby reinventing the wheel, forcing users to learn an unfamiliar set of phrasings and descriptive terms, and insuring that catalog records will be difficult to share and preserve in the long term–the goal is to take advantage of the benefits that standards extend both to creators and users of descriptive metadata.

With that in mind, a small group of EMP catalogers researched descriptive guidelines and standards that were used in cultural heritage repositories at the time and prepared an in-house cataloging manual. The standards upon which we ultimately based our own descriptions, with some exceptions that will be discussed below, were:

- *Anglo-American Cataloguing Rules*. 2d ed.[3]
- *Graphic Materials: Rules for Describing Original Items and Historical Collections*.[4]
- *Cataloging Internet Resources: A Manual and Practical Guide*.[5]
- *CONSER Cataloging Manual*.[6]
- *Archives, Personal Papers, and Manuscripts: A Cataloging Manual for Archival Repositories, Historical Societies, and Manuscript Libraries*.[7]

- *International Guidelines for Museum Object Information: The CIDOC Information Categories.*[8]
- *Museum Handbook.*[9]
- *Archival Moving Image Materials: A Cataloging Manual.*[10]

We also followed *The Chicago Manual of Style* in matters related to grammar and citations included in notes, and *The Associated Press Stylebook and Libel Manual* on capitalization and punctuation.

Most of the elements in an EMP catalog record conform to the areas of description outlined in *AACR2* and the special materials rules: Title and Statement of Responsibility, Edition, Publication Statement, Physical Description, Series, and various Notes. For most of these elements, the cataloging rules[11] listed above, when incorporated into the EMP cataloging manual, provide all the guidance needed for the catalogers to prepare descriptive data in the cataloging database. There are instances, however, when the rules have been adjusted somewhat due to either the nature of the materials themselves or the uses to which the information would be put in the public version of the EMP Digital Collection. Some specific examples illustrate how standard cataloging conventions were occasionally modified for local use.

In an EMP catalog record, the Title element plays a crucial role. It is displayed in the catalog's Title field, of course, but title information also forms the basis for captions used in galleries and online exhibits. It is the only information shown in the initial brief display of a retrieval set in the online Digital Collection and in the onsite Digital Lab. The Title element must therefore function as a sort of micro version of a full EMP catalog record, so slightly more information is included in devised titles than standard cataloging rules generally permit.

For published materials, EMP catalogers (following *AACR2*) transcribe the title from the chief source of information. Because of the nature of EMP's collection, however, many items, such as unpublished or ephemeral texts, graphic materials, and three-dimensional artifacts, have no formal titles. Cataloging rules and museum guidelines, especially Archives, Personal Papers, and Manuscripts and *Graphic Materials*, provide helpful directions on supplying titles for such materials, but we have combined and enhanced these rules somewhat for EMP cataloging purposes. A devised title in an EMP record includes, as appropriate:

> Main person or group (WHO) or thing or event or action (WHAT), possible location (WHERE), date (WHEN), and possibly statement of significance (WHY).

Some of the elements included in an EMP devised title are meant not only to identify the item but to convey its significance, specifically, the "when" and "why" elements. The date of creation, manufacture, or capture, for example, might serve to distinguish similar objects from one another and communicate something about an item's meaning or importance. For example, there is a major difference between a poster of Jimi Hendrix created or published in 1968 and one created in 1998.

The date of a performance affiliated with a poster or photograph may also provide the catalog user with clues as to the artifact's significance. A poster that advertised a performance by Nirvana prior to September of 1991, when their breakthrough album *Nevermind* was released, is significantly different from a poster promoting a performance in 1993, when they were established stars. Place of performance is also an important indicator of significance: in this case the older poster would have promoted a Nirvana performance in a small club, whereas the 1993 poster most likely would have promoted a performance in a large arena. For these reasons, catalogers are instructed to provide in all posters promoting performance events the names of the performers, the venue and city, and the date. For example:

> Mudhoney, Fastbacks, Alice in Chains, at the Moore Theatre, Seattle, WA, June 30, 1992

Historical or cultural significance is generally explained in a note element labeled History Remarks, but because an item may be in the EMP collection solely because of the role it played in shaping rock history, the Title element may also include, in addition to name and creation date, briefly stated information on an object's significance. For example, an object might be acquired because it was previously owned or used by a particular artist. This is usually indicated by adding "formerly owned by [name]" as a subtitle:

> Fender Stratocaster, 1968: formerly owned by Jimi Hendrix

> Adidas Superstars sneakers, circa 1985: formerly owned by Darryl McDaniels

In the case of the Fender Stratocaster, the History Remarks further explain that this is the guitar Hendrix played in his legendary performance at the Woodstock rock festival.

The History Remarks note was developed by EMP's cataloging staff. Although related to the Scope/Content note, which describes the intellectual content of an item, the History Remarks note places the item in its historical or cultural context. It explains the significance of the artifact, describing its association with a person, a group of people, an era, a place, or a type of music. The text is generally developed from information found in secondary sources, such as books, Web sites, museum acquisition files, and discussion with curators. Sometimes, History Remarks are taken from text created for purposes other than cataloging. For example, during the late 1990s EMP's Web site included a feature called The Vault, in which each day a different artifact was digitally displayed and described in a brief historical essay written principally by curatorial and cataloging staff members. When The Vault was discontinued, each essay was included in or linked to the History Remarks note in the related artifact's catalog record.

Catalogers create History Remarks using EMP's Oral History Program transcripts, which include interviews with both well-known and behind-the-scenes figures in popular music. Since many of the artifacts in the museum's collection are sold, loaned, or donated by historically significant musicians, songwriters, producers, designers, record label executives and others, EMP's curators frequently schedule oral history interviews during the acquisition process. In addition to capturing the subjects' narratives of their lives and their experiences in the music industry, the curator also asks interviewees to describe the origin and significance of the artifact being acquired. As a result, catalogers writing History Remarks for these items can often include an exclusive, first-hand account of the object provided by the source of its significance.

Only high priority artifacts (defined as items intended for exhibition in the museum or on the Web) are researched and described in the History Remarks field; nonetheless, preparing these notes takes time and effort. When asked why EMP took this labor-intensive approach to cataloging, curator Jim Fricke explained, "The importance of contextual information in the records presented itself as obvious: many of the 'artifacts' that we collect are mass-produced objects that require such information to establish their significance and distinguish them from otherwise identical objects . . . The point of information is to inform. My goal in cataloging is not just to describe an object, but to convey why we felt is was worth preserving."[12]

Telling the stories related to culturally and historically significant artifacts is a critical component of their presentation. More meaning is conveyed when a drinking cup exhibited in an ethnographic museum is accompanied by an explanation of its ceremonial use, or a handwritten letter displayed in a history museum is revealed to have underlying significance in the writer's life. Similarly, an outfit worn on stage or a musical instrument that played a part in the history of American popular music is far more meaningful to the museum visitor when that history is described. Why did EMP acquire the white Fender Stratocaster guitar featured both on the Web site and in the Jimi Hendrix gallery? The note in the Historical Remarks field explains it:

> *History remarks*: This is the Fender Stratocaster that Jimi Hendrix played during his legendary performance at the Woodstock Music and Arts Fair in upstate New York on August 18, 1969, when he created one of the most famous musical moments of the 1960s with his rendition of the "Star Spangled Banner." Hendrix used this guitar for many performances between October 1968 and September 1970.
>
> Hendrix purchased the guitar at Manny's Music in New York City, probably in the late summer of 1968. He may have personally made the necessary adjustments to make it a left-handed instrument, including reversing the strings, replacing the nut in order to facilitate the reversed strings, and switching the position of the strap pin on the fretboard side from the longer horn to the shorter horn.

Another cataloging requirement unique to EMP is the need to describe and uniquely identify items consisting of multiple parts. A curator might decide to exhibit one disc included in a box set comprised of three discs, a specially designed container, and a booklet. Because any piece of a multi-part artifact can be exhibited, each part must be numbered, tracked, described, and indexed separately, but at the same time, it is important to link each part to the information pertaining to the artifact as a whole. The concept of multi-level description employed by archivists to describe large collections at successively narrower levels of specificity[13] was a useful model for us and formed the basis of the hierarchical structure used to catalog multi-part objects in EMP's catalog.

NAME INDEXING

The instructions for providing access points to names are, like EMP's descriptive cataloging rules, based on *AACR2*. Personal and corporate names are indexed as "Association" elements in EMP artifact records; these are the equivalent of personal or corporate names indexed as creators (1XX and 7XX) or subjects (6XX) of an item in an *AACR2*/MARC record. (See Figure 1 for sample input window.) EMP indexes every name associated with an artifact, including publishers and printers of books, and designers and manufacturers of clothing and other three-dimensional objects. The museum is interested in the "behind-the-scenes" people involved in the creation and manufacture of collection artifacts, and because exhibits may be built around these lesser-known figures, it is important that they be indexed.

"Role" terms showing the connection between a person or organization and the item being cataloged are included in Association entries. Almost all of the role terms used in EMP artifact records are taken from the *MARC Code Lists for Relators, Sources, Description Conventions*

FIGURE 1. Basic CIS Cataloging Screen. Clicking on any item in the boxes on the right side of the screen opens a new input window. E.g., to index names, the cataloger clicks on "Associations," causing the name indexing input screen to open.

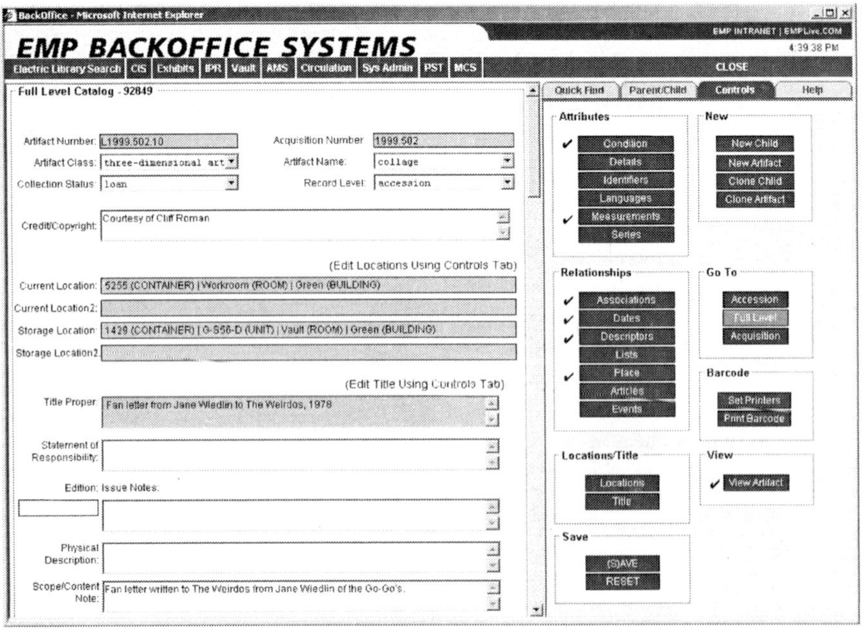

(http://www.loc.gov/marc/relators/relaterm.html). Here are some typical roles indexed for various types of artifacts:

- *Graphic items*: Subject (such as the performers and venues of an event being promoted on a poster or handbill), visual image (i.e., who is pictured on an item), designer, artist, printer, publisher, photographer (for any photographic images on an item), and copyright holder;
- *Sound recordings*: Performing artist, producer, record company, distributor, and copyright holder;
- *Printed materials*: Author, publisher, editor, visual image (if pictured on the front cover), subject, and printer;
- *Three-dimensional items*: Manufacturer and/or creator.

There are three Role terms that may be applicable to all artifacts: "former owner," "signer" (someone who personally autographed an item), and "associated name." Also taken from the MARC relator term list, "associated name" is a catch-all for names which are not connected to an item in a specific way, but still need to be indexed. An example of a name receiving an "associated name" role term would be "Jimi Hendrix" for a "Jimi Hendrix signature" guitar manufactured by Fender. Hendrix was not involved with the design and creation of the guitar and he didn't actually sign it, but because his name is associated with the instrument, it is indexed as an "associated name."

NAME AUTHORITY FILE

The people associated with artifacts are as important as the artifacts themselves at EMP. Curators frequently write biographical sketches and band genealogies for the galleries and EMP Web site, and they often do research in the in-house name authority records created by the cataloging staff. As with traditional library authority records, EMP name authority records include name variants as well as earlier and later name forms. Unlike traditional library authority records that seek only to uniquely identify personal and corporate names, full EMP name authority records provide fairly detailed biographical notes, band membership listings, and descriptors for instruments and types of music played. (See Figure 2 for sample name authority input screen.) As with the History Remarks field, full information is provided only for high-priority personal and corporate names (i.e., names featured promi-

FIGURE 2. CIS Name Authority Input Screen. Clicking on the Relationships button opens additional screens for recording various forms of the name, dates, descriptors, and related personal and corporate names such as band member names.

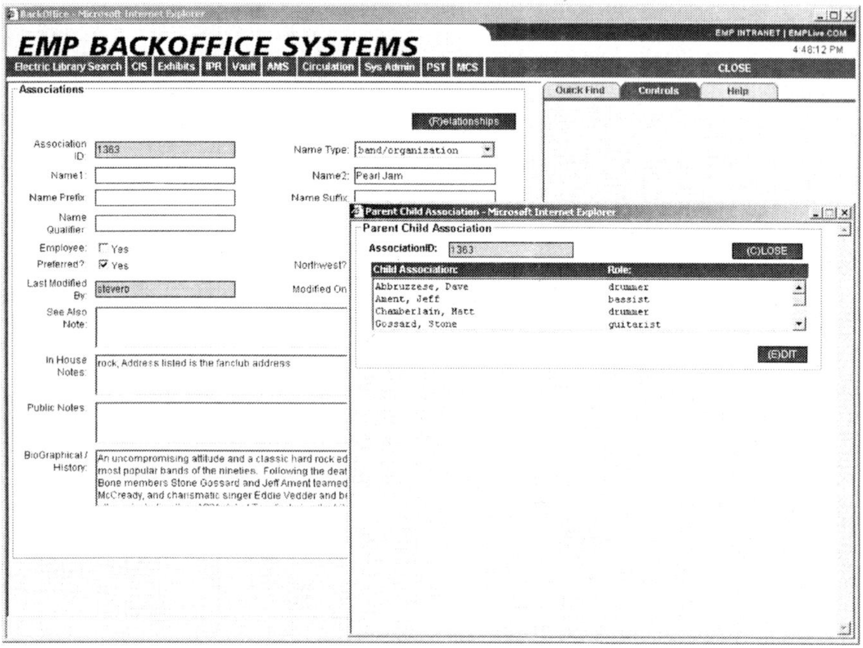

nently in the museum galleries and exhibits or associated with exhibited artifacts).

DESCRIPTORS: PROVIDING SUBJECT AND FORM/GENRE ACCESS

Because CIS is a catalog of specialized museum artifacts, EMP catalogers typically provide more descriptors for an item than would be found in a traditional library catalog record. Just as in a MARC record (in the 6XX fields), several "descriptor types" may be assigned.

- *Form/genre*: The form/genre descriptor type is used if this aspect of the item is not evident from the artifact name. For example, a poster originally created to advertise a performance would be indexed with the form/genre descriptor "promotional materials." The form/genre descriptor type may also be used to bring out an aspect of an artifact that is not inherently part of the item. For example, a guitar signed by the members of Nirvana would be indexed with the form/genre descriptor "autographs (signatures)."

- *Style*: The style descriptor type is used to index musical or artistic style. Examples include "rock (music)" or "jazz (music)" for a sound recording, "psychedelic (style)" or "boxing style (graphic design)" for a poster, or "Western (style)" or "hip-hop (style)" for an article of clothing.
- *Visual image*: The visual image descriptor type is used to index what is pictured on an item, so generally this is used for graphic materials such as posters and photographs.
- *Topical*: Topical descriptor types correspond to the traditional library subject heading or "aboutness" of an item. Although it may be applied to other types of artifacts, this is most often used for text-based print items like books and magazines.

EMP THESAURUS

The locally created and maintained EMP thesaurus, which utilizes a commercially available thesaurus construction package, contains the controlled vocabulary for three fields within artifact records. The first of these is Artifact Name, which refers to the generic term for an item, such as "album (sound recording)," "poster," "electric guitar," etc. This field is essentially the equivalent of the Specific Material Designation in a MARC record. The thesaurus is also the source of topical descriptors used in EMP cataloging, and it includes Association role terms as well.

Primarily based on ANSI/NISO Z39.19, *Guidelines for the Construction, Format, and Management of Monolingual Thesauri*, the EMP thesaurus is organized into several major hierarchies: Activities, Agents, Animals (used primarily for animal images depicted in visual arts materials, such as posters and handbills), Associated Concepts, Materials, Objects, and Styles. The three most developed hierarchies are Agents (used for the Association role field and descriptors related to performing artists), Objects (used for the Artifact Name terms), and Styles (used for musical and artistic styles).

A large percentage of the terms found in the thesaurus are taken from the Getty Research Institute's *Art and Architecture Thesaurus* (http://www.getty.edu/research/conducting_research/vocabularies/aat/index.html).

Because each entry includes all necessary information, including references and scope notes, *AAT* terms are usually incorporated into the EMP thesaurus unaltered. Of course, the *AAT* is more art than music-oriented, so additional information sources are needed for terms included in EMP's thesaurus.

The *Library of Congress Subject Headings* list is used as a secondary source of terms, although LCSH is still a little too general for EMP. When necessary, the museum's thesaurus administrator enhances LC terms with local scope and usage notes through consultation of general and specialized dictionaries and other reference materials. Some EMP terms are not found in any of the previously mentioned sources, so more specialized sources are reviewed, including books about popular culture and music, fashion, music technology, graphic arts, and other topics represented in the collections.

Unfortunately, even with a multitude of sources available for consultation, there are still terms for which information is unavailable, and in those instances a staff member must invent a term and its definition. Terms used in a catalog record's Artifact Name field, such as "admission ticket book," and "drumstick holder," are examples of original terms. The definition of a particular musical style term may be ambiguous and even controversial, and requires the expertise of curatorial and other EMP staff to clearly establish local usage.

As EMP's collections grow, the thesaurus used to index these new artifacts continues to evolve. The chief cataloger is in charge of maintaining and expanding the thesaurus. New terms are proposed by catalogers and curators, and added to the thesaurus if approved by cataloging staff at the regularly held vocabulary review meeting. Visitors to the EMP Digital Lab or the Web-based EMP Digital Collection (http://www.emplive.com/digitalcollection/index.asp) are viewing terms included in the EMP thesaurus when they browse descriptors listed in the "Styles" and "Things" categories, or when they click on an artifact's "Related Keywords." The thesaurus itself, however, is not available to the public at this time.

DATABASE STRUCTURE

Although we've emphasized the cataloging staff's resolve to build the EMP catalog on existing descriptive standards, we were simply not able to prepare records in MARC format. The museum's artifacts are described to a finer level of granularity than MARC accommodates, and

MARC is not an effective structure for the hierarchical records we create for multi-part artifacts. The EMP catalog contains many non-public fields, such as acquisitions, insurance, loan, and detailed physical description and condition reports that would have had to be fitted into non-specific or other inappropriate MARC fields. In addition, the need to link hierarchical catalog records to associated digital reproduction files (sometimes multiple digital files) made MARC an unsuitable structural standard on which to base the EMP catalog.

In fact, it was clear early on that no off-the-shelf database package would serve the myriad functions of the Collections Information System as envisioned by the staff. A succession of EMP curators, catalogers, programmers, and consultants helped develop the database from its original flat-file structure to the present far more complex relational scheme presently used in SQL Server. The current database structure and data entry interface was finalized in 1999 and has been used ever since. Catalogers enter information into a non-public, Web-based descriptive cataloging screen that contains linked screens for adding Associations (personal and corporate names and role terms) and Descriptors (taken from the EMP thesaurus), as well as screens for describing components of an artifact (along with their names and descriptors) as needed.

DIGITAL REPRODUCTION DATABASE

As CIS was evolving, the digital photography staff systematically created high-quality digital reproductions of every artifact selected for exhibition in the museum galleries and on the Web site. Each component of a multi-part artifact was individually photographed, and some three-dimensional objects were shot at various angles and on various sides. High-resolution TIFF files were made from the original proprietary image files created by a digital camera, so there were several versions of each image. In addition to the archival and display versions of these digital image files, sound recordings and moving images were digitized.[14]

With digital reproduction files growing exponentially, an off-the-shelf digital asset management package was purchased in the late 1990s to allow staff to identify, track, and provide administrative and technical metadata for the files. Although the cataloging staff submitted recommendations on metadata elements to include in the digital object database, it was primarily staff members in automated systems and the digital audio, video, and photography labs who made these decisions.

Administrative and technical metadata provided about digital files, available for in-house use only, includes such information as date of capture, sampling frequency, file duration/size, resolution, compression information, and so forth. Once the digital asset management database was up and running, a staff member was hired to track and provide technical metadata for EMP's digital object files.

By 1999, CIS functioned as the artifact cataloging database, and the asset management package identified related digital reproduction files; what was missing was a link between the two databases. Creating a seamless link between a complex relational cataloging database (one that also contained hierarchical relationships) and a digital asset management system containing thousands of files in multiple formats was far from trivial. A prototype appeared fairly quickly, but it took several months and a lot of creative programming before the in-house version of the system was ready for day-to-day use.

Once the fully functioning staff version was in place, and catalogers could view not only the original artifacts but also their digital surrogates while creating catalog records, an outside Web design firm was engaged to provide a public interface to the system. The public search and display interface needed to combine powerful access features with the vibrant sense of design that distinguishes every aspect of EMP, from gallery presentations to studios and performance venues to online exhibits, and it needed to be accessible both in workstations inside the museum and on the World Wide Web. The designers worked their magic, and by opening day, the Digital Lab was ready.

THE EMP DIGITAL LAB

The Digital Lab is an onsite facility that provides public access to a large selection of the museum's holdings. The Digital Lab database features four navigational tools useful for browsing and searching the available artifacts (most, but not all of these tools are also available on the Web-based EMP Digital Collection, http://www.emplive.com/digitalcollection/index.asp):

- *Mixer*: a browsing function that allows the user to progressively filter the artifacts currently available in the public catalog. Under the categories Names, Styles, Things, and Places, each of which represents an index of names or terms, a user can click on descriptors to access artifacts of interest. The user may narrow the initial browsing results by clicking on additional categories. For example, selecting "blues" from the Style category retrieves several hundred artifacts. By next selecting "album (sound recording)" from the Things category, the retrieval set is narrowed to a sorted list of blues albums. The user can then click on an album title for a full-level display of cataloging information, digital views of both sides of the disc and its jacket, and full-length audio tracks.

- *Gallery Views*: a tool displaying artifacts by their location in the onsite exhibits. For a new visitor, this section provides a preview of what EMP has to offer, while those already familiar with the galleries can explore areas of particular interest in greater depth.
- *My Bookmarks*: an electronic artifact bookmarking system for visitors to the museum building in Seattle who have used the MEG™ (Museum Exhibit Guide). A hand-operated audio device that plays hours of exhibit-related stories, commentary, and song tracks on demand, the MEG™ also enables each visitor to bookmark items of personal interest while browsing the galleries. These customized lists, linked to the user's ticket ID number, are accessible in both the onsite and Web versions of the Digital Collection, so museum guests can spend more time learning about favorite artifacts after their gallery visit ends.
- *Search*: a full-text search feature available only in the onsite Digital Lab. The user may search the full text of the public catalog record fields.

For each artifact retrieved via the browse or search function, the user sees selected fields from the artifact catalog record: title, object name, statement of responsibility, creation/release dates, historical background note, and indexed names and topical headings. (Each of the name and topical headings is hyperlinked to records for all other artifacts indexed under those headings, enabling navigation to related items.) Clicking on "Details" provides additional information: place of creation/publication, artifact ID numbers, and names of component parts if the artifact consists of several parts. Clicking on a component part retrieves additional information about that component (such as the author of liner notes in a booklet included as part of a CD set). Descriptive elements not visible to the public include administrative information such as detailed physical description and condition notes. (See Figure 3.)

Also visible are thumbnail-sized digital image(s) of the artifact, which usually provide several views of the item (both sides of a 45 rpm record, for example). Clicking on a thumbnail opens the "image viewer," which

FIGURE 3. Basic EMP Digital Collection Record

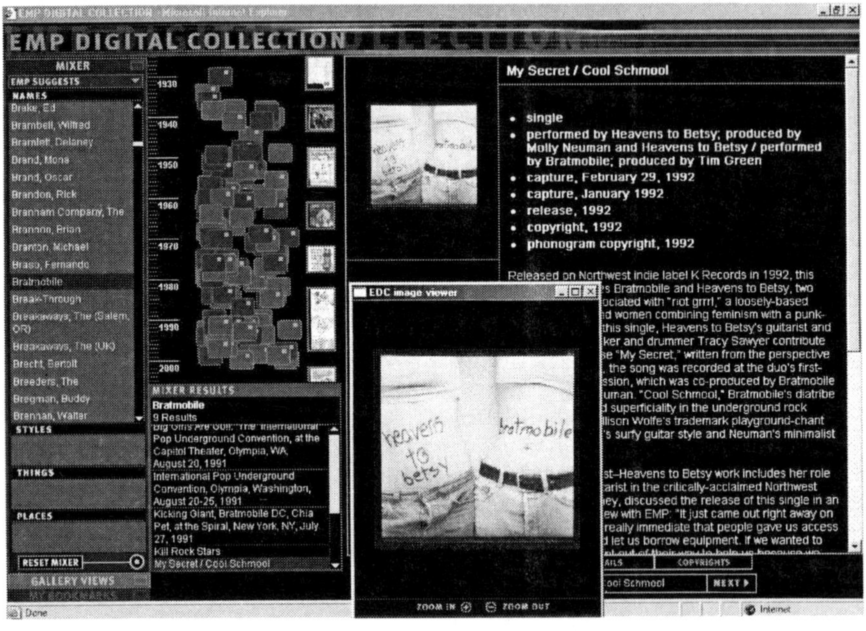

Reprinted with permission.

shows an enlarged version of the image and allows the user to zoom in or out. In the Digital Lab at the museum, users may also launch an audio player to hear digital audio of the item or a video player to see a moving image item. For copyright reasons, only digital still image files and brief audio clips (usually just under 30 seconds in length) are available on the Web site.

DIGITAL LAB USER SURVEY RESULTS

Like EMP as a whole, the Digital Lab is a popular success, with attendance figures showing that over 13% of museum visitors stopped by to explore the virtual collection by the end of the year. This figure has steadily increased, reaching 17% by the end of 2002, in part because of the work of EMP's Education staff to promote the Lab as a valuable educational tool for school group tours. Attendance has been further boosted by technological enhancements to the Digital Collection database, such as the inclusion in 2001 of the museum's gallery films (incorporating video clips from EMP's extensive oral history program, along with concert performance footage) and the addition of full-text search functionality to the user interface in 2002.

The Digital Lab also plays a part in the promotional efforts surrounding the opening of new exhibits in the museum's temporary gallery. As part of EMP's goal to provide an ever-changing experience for visitors, members of the cataloging staff conduct research and compose historical background notes about the new artifacts, incorporating these into full-level cataloging records added to the database for each new item on display.

To study user response to these efforts, the authors conducted a survey of Digital Lab visitors throughout June 2003, collecting a total of 212 questionnaires representing a sample of 3.9% of the number of visitors to the museum that month. Most respondents (84 people, or 39.62%) reported a visit of 15 to 30 minutes in duration to explore the Digital Collection, while 31.6% stayed for shorter and 24.1% for longer sessions. In general, the responses suggested that users were satisfied with both the informational content and the organization of the database. The figures below represent the number of users who selected "Strongly agree" or "Agree" to the following statements:

- Searching the Digital Collection was easy: 60.4%
- I was able to find the artifacts I'm interested in: 63.7%
- I was satisfied with the content of the Digital Collection: 74.5%
- I was satisfied with the presentation of the Digital Collection: 84.9%

Paradoxically, while the presentation of cataloging records in the database was among the highest-scoring features, it also attracted the greatest number of negative comments in the open-ended section of our survey. By far the most frequent response to the request for suggestions on improving the Digital Collection were those that focused on the user interface, with 35 of the 89 respondents who provided comments mentioning at least initial difficulties with the search mechanisms. Typical of these were the following: "I'm computer literate, but it took me a while to figure out how to search, needs to be less complex"; "Very slick but not obvious how to do things"; "More intuitive interface–very confusing to a casual audience; looks great, good idea, but [it has a] bad set up."

Designed by an outside contractor, the museum build of EMP's Digital Collection database features a dynamic interface that displays result sets graphically. Browsable links based on shared data elements (such as descriptors, name headings and artifact types) in the cataloging records connect similar items, in an attempt to visually represent the inter-relatedness and influences among popular music subgenres and performing artists through their corresponding artifacts in EMP's collection. While many visitors respond enthusiastically to this cutting-edge technology, museum staff members have long noted that others, especially those less experienced with computers, can find it frustrating. This situation has been remedied somewhat by the implementation of a more traditional text-based search mechanism developed and added to the database by EMP's in-house technology department in 2002. However, according to our survey results, some users were unable to even locate this feature in the current interface. As EMP develops from a start-up museum into a mature institution, it is hoped that further redesigns incorporating more cataloger input and results of in-depth user studies can help remedy this occasional disconnect between the catalog record and its user.

Despite some limitations, the Digital Lab plays an important role in EMP's mission to provide a dynamic, interactive, and educational visit. From the full-length audio tracks linked to all sound recording catalog records, to the archive of past temporary exhibits featuring artifacts no longer on display in the building, the Digital Collection attempts to transcend the traditional physical constraints of the museum walls to provide an enhanced visitor experience. This attempt is often successful, according to our survey. Some of the survey comments included: "This blew me away–tremendous!"; "Truly interactive, really brings the whole experience alive"; "A great addition to EMP–the best bit of the museum!" Perhaps for these users, and for the catalogers who provided the work that inspired this response, the search for nirvana has become a little more attainable.

NOTES

1. Fricke, Jim. "Proposal for a Collections Information System," in-house memo, 1995.

2. Heery, Rachel, and Manjula Patel, "Application Profiles: Mixing and Matching Metadata Schemas," *Ariadne*, issue 25 (Sept. 2000). [Accessed 14 January 2004] Available at URL: http://www.ariadne.ac.uk/issue25/app-profiles/.

3. *Anglo-American Cataloguing Rules*. 2d ed. Chicago: American Library Association, 2002-.

4. Betz, Elisabeth W. *Graphic Materials: Rules for Describing Original Items and Historical Collections.* Washington, DC: Library of Congress, 1982. Available with updates through 1997 at URL: http://lcweb.loc.gov/rr/print/gm/graphmat.html.

5. *Cataloging Internet Resources: A Manual and Practical Guide.* 2nd ed. Nancy B. Olson, ed. Dublin, Ohio: OCLC, 1997.

6. *CONSER Cataloging Manual.* Jean L. Hirons, ed. Washington, DC: Library of Congress, Cataloging Distribution Service, 2002-.

7. Hensen, Steven L. *Archives, Personal Papers, and Manuscripts: A Cataloging Manual for Archival Repositories, Historical Societies, and Manuscript Libraries.* 2nd ed. Chicago: Society of American Archivists, 1989.

8. *International Guidelines for Museum Object Information: The CIDOC Information Categories.* Developed by the International Committee for Documentation (CIDOC) of the International Council of Museums (ICOM). Paris: ICOM, 1995. Printed and online versions available. URL: http://www.willpowerinfo.myby.co.uk/cidoc/guide/guide.htm.

9. National Park Service. *Museum Handbook.* Washington, DC: U.S. Dept. of Interior, 1990-. Part II provides guidance on accessioning and cataloging. Looseleaf printed and online versions available. URL: http://www.cr.nps.gov/museum/publications/handbook.html.

10. *Archival Moving Image Materials: A Cataloging Manual.* AMIM Revision Committee; Motion Picture, Broadcasting, and Recorded Sound Division. Washington, DC: Library of Congress, Cataloging Distribution Service, 2000. Printed and online versions available. URL: http://www.tlcdelivers.com/tlc/crs/arch0332.htm.

11. The *CIDOC Categories*, while more properly described as a structural standard, also provided useful guidance on content values.

12. E-mail from Jim Fricke, Dec. 1, 2003.

13. See Principle 3 in the "Statement of Principles for the CUSTARD Project" (Canadian-U.S. Task Force on Archival Description), Society of American Archivists Web site. [Accessed 14 January 2004] Available at URL: http://www.archivists.org/news/custardproject.asp.

14. Textual items such as typed and handwritten song lyrics were digitally photographed, but most of these have not as yet been converted to searchable text files via optical character recognition (OCR) software.

Collaborative Cataloging: Using Dublin Core to Unite Local Heritage Organizations

Kody Janney

SUMMARY. Publishing images online is challenging enough for any one institution. When the University of Washington Libraries and the Museum of History & Industry, Seattle decided to do so and include ten small local heritage organizations in the process, Dublin Core provided the glue that made it work. The goal of providing integrated Web access to the combined materials from all the museums, while maintaining individualized databases for each organization was achieved through the use of standards, hard-work, cooperation, and an IMLS grant. *[Article copies available for a fee from The Haworth Document Delivery Service: 1-800-HAWORTH. E-mail address: <docdelivery@haworthpress.com> Website: <http://www.HaworthPress.com> © 2004 by The Haworth Press, Inc. All rights reserved.]*

Kody Janney, MLIS, is the proprietor of KJ Consulting, an information management consulting company. She was formerly Coordinator of Digital Initiatives for the University of Washington Libraries and Co-Principal Investigator of the IMLS grant discussed in this article (E-mail: kjconsul@nwlink.com).

The author would like to thank Rachel Howard, KC Snapshots Project Manager, and Mary Montgomery, Co-Principal Investigator, both of the Museum of History & Industry, Seattle. This article would not have been possible without them.

[Haworth co-indexing entry note]: "Collaborative Cataloging: Using Dublin Core to Unite Local Heritage Organizations." Janney, Kody. Co-published simultaneously in *Journal of Internet Cataloging* (The Haworth Information Press, an imprint of The Haworth Press, Inc.) Vol. 7, No. 1, 2004, pp. 33-48; and: *Collaborative Access to Virtual Museum Collection Information: Seeing Through the Walls* (ed: Bernadette G. Callery) The Haworth Information Press, an imprint of The Haworth Press, Inc., 2004, pp. 33-48. Single or multiple copies of this article are available for a fee from The Haworth Document Delivery Service [1-800-HAWORTH, 9:00 a.m. - 5:00 p.m. (EST). E-mail address: docdelivery@haworthpress.com].

http://www.haworthpress.com/web/JIC

Digital Object Identifier: 10.1300/J141v07n01_03

KEYWORDS. Dublin Core, CONTENTdm software, metadata, data dictionaries, image databases, historical photographs, field mapping

PROJECT BACKGROUND

In 2001, the Institute for Museum and Library Services (IMLS) awarded the Museum of History & Industry (MOHAI) in Seattle, Washington and the University of Washington Libraries (UW) a National Leadership Grant for Library-Museum Collaboration. The grant provided for these two institutions to lead ten local heritage museums and societies step-by-step through the selection, scanning, cataloging, and Web presentation of a total of twelve thousand historical images of King County, Washington from their combined collections. Grant funding paid for scanning services by an outside contractor and for project staff salaries. Professional staff from MOHAI and UW trained project partners in collections management, cataloging, and computer software. Metadata entry was performed by two paid staff members hired by the project, with supplemental work performed by the staff of some of the organizations. The end result of the grant was separate image databases for each participant and the project's Web site, King County Snapshots (KC Snapshots) available at http://ww.kcsnapshots.org, from which users have the ability to search across all the databases. A complete list of project participants is listed in the Resources section at the end of this article.

The University of Washington Libraries, one of the largest research libraries in the United States, already had twenty-five thousand cataloged images available online, and had the software, servers, expertise, and willingness to incorporate databases of images from partner institutions.

The Museum of History & Industry is a mid-sized urban museum, with a staff of about twenty. Its photo collection contains over 1.5 million items. Many of these photos were donated to MOHAI by the Seattle Post-Intelligencer newspaper. About two thousand of their images were already accessible online. Images reside in MOHAI's library which is staffed by a professional librarian.

MOHAI has a healthy relationship with smaller museums and historical societies in King County, Washington, and knew that they too had important underutilized image collections. These small heritage organizations also have chronic funding and staffing shortages. Many of them are run entirely by volunteers. Some have no public hours, and access to collection materials is limited at best. They–and the communities they serve–would benefit from online indexed images, but without robust Web sites, in-house skills, or the time to write grants, the task seemed impossible. Paid staffing at these museums averages three full-time employees per organization, and three institutions have no paid staff whatsoever. Budgets are extremely limited; the median annual budget for the ten smaller participants is eighty thousand dollars. The largest number of images that any of these smaller organizations had online prior to the project was sixty, but the photographic holdings per institution is at least a thousand, with quite a few having well over that number. These smaller museums have primarily had community volunteers, not librarians, archivists, or curators in charge of their collections. Many of them have never had a trained professional on staff in their history. As a consequence, accessioning and cataloging of the images was usually spotty and almost always involved multiple disparate systems within each institution.

The project would have been impossible without CONTENTdm, a Web-based digital asset management system (http://www.contentdm.com). In addition to providing internal tools for managing digital materials, this software provides mechanisms for publishing and searching digital collections over the Internet. Each project participant's images and descriptive metadata are stored in a CONTENTdm database on UW's server. Dublin Core (DC) serves as a crosswalk behind local field names connecting the different databases. Each organization can provide the ability to search its holdings from its own home page, and the Web site for the entire project allows unified searching across all collections. The success of these searches depends on the uniformity of cataloging and the use of Dublin Core by all participants.

THE PROCESS OF CREATING THE KC SNAPSHOTS DATABASE

Project leads at each organization committed themselves and their organizations to the project and attended three training sessions hosted by MOHAI and the UW Libraries. One of these sessions was devoted to learning about metadata and creating it hands-on using the project's software. In addition, the project metadata specialists, Theodore Gerontakos and Julie Kerssen, did follow-up coaching for the project leads in the use of CONTENTdm and cataloging. As the metadata specialists created minimal-level or higher-level catalog records, depending on the amount of information supplied, they created a list of questions to be answered by each museum's project leads, who were given read/write access to their organization's database so they could apply the answers directly to the metadata. The specialists then moved on to the next organization, but continued to follow up with each group. Most of the minimum-level cataloging was performed at the individual institutions, although technological issues, such as firewalls and inadequate Internet access, required some of it to be done elsewhere or created off-line in Excel to be loaded into the system later.

Each project metadata specialist was assigned to specific smaller organizations, and the two of them shared the responsibility for cataloging the images from the lead partners. They spent time at their assigned organizations, and worked with staff and volunteers to gather information about the images. Many of the organizations use PastPerfect museum collection management software. In these cases, and other instances of computerized museum collection information, data was exported as ASCII text into Microsoft Excel spreadsheets. These were then imported into the CONTENTdm software, and associated with the appropriate images to seed the image database with metadata in the correct fields.

At every institution, image collections are cataloged–if they are cataloged at all–using a variety of schemes. Some of these methods were inherited when the museum obtained an image collection. Some were cataloged using systems that were popular at the time the material was acquired or using methods that were invented by a particular registrar, but have since been superseded. Much as they would like to, the project partners have not been able to do retrospective cataloging. Consequently information was garnered from scrawls on scraps of paper, writing on image versos, or handwritten lists, and transcribed into *DC Description* and other appropriate fields, with sources noted. Institutional and local memories were mined for factual, contextual, and anecdotal information by the metadata specialists and by the various organizations' memberships as they worked on the project.

The metadata specialists reported to Rachel Howard, the project manager, and remained in regular contact with her and with each other. The project manager reviewed the work of the metadata specialists, discussed any inconsistencies in the records, and ensured that appropriate changes were made. This interaction helped ensure consistency in metadata terminology and style across organizations. For example, *b&w* is used throughout for black and white images, per Anglo-American Cataloging Rules, Second Edition (AACR2), as opposed to the many other possible variations such as *black and white*, *black & white*, and *b & w*. Another example is the use of *n.d.* for no date, which was decided upon by the group, instead of variations such as *date unknown*, *no date*, *undetermined date*, or completely omitting date information in cases where it is unknown. The project manager was also able to create more consistent *Description* field styles between the two metadata specialists.

Because of the training, project leads at the various organizations will be able to continue enhancing the metadata in the future. The University of Washington Libraries will continue hosting the CONTENTdm software as part of their Digital Initiatives program.

DATA DICTIONARIES

Cataloging to mutually agreed-upon standards would have been difficult if the grant had not provided for hiring the two metadata specialists, and if the University of Washington Libraries had not had a Metadata Implementation Group (MIG). This standing committee of catalogers and other information specialists within the UW library system develops and promotes the use of metadata standards to ensure reliable resource discovery within and across UW's digital collections. When reviewing data dictionaries for digital collections, MIG also helps determine appropriate controlled vocabularies. MIG, with Kris Kinsey as project liaison, chose Library of Congress (LC) Name Authorities and the Thesaurus for Graphic Materials (TGM) for this project.

Guided by this expertise, the project's metadata specialists drafted data dictionaries for each partner organization and submitted them to UW's Metadata Implementation Group for approval before cataloging began. The partners discussed their existing and desired access points, and the metadata specialists mapped those to Dublin Core. While it would have been ideal for all organizations involved in the project to use the same data dictionary, existing cataloging and the various focus points of each organization made this impractical. Even the concept of a data dictionary with standardized fields across all material within a single organization was revolutionary for some of the groups. MIG was also consulted on other issues as they arose.

To process its KC Snapshots material, MIG recommended that UW use the standard data dictionary it had developed over several years for the historical photograph collections in the Libraries' Manuscripts and

Special Collections branch. This ensured that the new material would be compatible with the other digital image collections UW had already created. It also gave the KC Snapshots project the benefit of their experience. This data dictionary was presented to MOHAI, which then, with the help of MIG, modified it for use by museums. This modified data dictionary was presented to all of the smaller organizations. While they were free to change any of the fields that were not required administratively, either by adding new ones or eliminating suggested ones, the advantages of uniformity and experience were stressed. Many of the museums made modifications, mostly small. Some of the modifications were only in the wording of a field's name, not the concept of the field or the type of values it would contain. For these small changes, Dublin Core mapping was not affected. Fields that could have been mapped to the *DC Subject* element were the ones that required the most changes, with increased specificity often being requested.

One of the issues that frequently came up when the data dictionaries were being designed was the desire of museums to put shelf location information into CONTENTdm. When it was explained that this would mean updating material in two places as there is no automatic interface between CONTENTdm and a collection management system, the disadvantages of this idea became clear and no one implemented it. One organization did ask for a *Shelf Location* field to make it easier to find the original photograph if it was requested for reprinting. Their material is not moved often, so this field should not require frequent updates.

CONTROLLED VOCABULARY

Facing reality and the differing nature of the missions of libraries and museums, MIG allowed local vocabularies as supplements to the authorized terms from the Library of Congress. Terms in the *Subjects* field are taken solely from approved terminology. Auxiliary vocabularies are listed in other fields, often with names like *Classification* or *Category*.

Three of the organizations already had their own in-house subject lists that they had used in their registration practice. The White River Valley Museum has a specific terminology called *Classification* which they developed in-house. Northwest Railway Museum has terminology from a nationally recognized railroading vocabulary. White River Valley decided to include a *Classification* field in their database. However, the Railway Museum concluded that they did not want to include their specialized list as well as the TGM terms in their records, and were content to use just the TGM terminology. Any detailed train terminology was put into the *Caption* or *Notes* field, as appropriate.

Two other organizations that had not previously had any vocabulary control and only limited subject access began creating subject terms or keywords on their own when they first started selecting images for the project. This was before they had the metadata training, and they did not know about the existing authorities. The project staff had no idea they were creating these in-house terminologies. When they were told about TGM, both museums decided to use the authorized vocabulary in lieu of their hastily constructed ones.

The Puget Sound Maritime Historical Society (PSMHS) had already indexed its vessel images on internal systems by the fields *Rig* and *Use.* These fields employ rich controlled vocabulary lists developed within the organization which use more specific terms than TGM has to offer. This terminology provided a depth of detail about the vessels that would have been lost by moving to the broader level TGM terms. PSMHS wanted to make sure that their members could access their images by these terms. However, the mixture of specific and more general terms in a unified database would mean that many potentially useful images would be missed if a *Rig* term was searched on the modified database, since other organizations were not interested in cataloging their maritime images to the same depth of detail as PSMHS.

Rig and *Use* often contain only contextual information about the ships shown, and not necessarily data about what is depicted in the image. The information contained in these two fields and the train terminology are the kind of information that other organizations were putting in their *Notes* or *Description* field. Because the grant's principal-investigators and the Metadata Implementation Group felt that the DC Subjects fields should not become diluted and, in essence, free-text, *Rig*, *Use*, and *Classification* were not mapped to DC Subjects. The DC Subjects fields were thereby limited to LC controlled vocabulary. Because White River Valley Museum's *Classification* field did reflect the content of the image, albeit at a high level of granularity, there was some discussion as to whether or not *Classification* should be mapped to *DC Subject.* In the end, consistent use of a nationally recognized controlled vocabulary won and the metadata specialist applied the appropriate terms to the images.

Neighborhood names are very important in local history, and needed to be included in the database. As the project dealt with many local names not found in LC Names Authorities, the Tacoma Public Library's Washington Place Names Database and Getty's Thesaurus of

Geographic Names were used as supplements. Neighborhoods that were not listed in LC were sent to the University of Washington Libraries' Metadata Implementation Group, which then made the necessary Library of Congress Subject Authority Cooperative Program (SACO) requests and the headings became official.

One partner, the Wing Luke Asian Museum, balked on learning that the LC authority term for their neighborhood, the subject of much of their collection, is "International District." Historically called "Chinatown," other Asian populations moving into the area had precipitated a name change; but that name change also became a political issue. The neighborhood's newest signage calls it "Chinatown/International District." The Wing Luke Asian Museum wanted to use this compromise name in its records. MIG took the proposed name change to the Library of Congress. Since the post office branch and the neighborhood library branch still used "International District," and some official documents did not use the new name yet, the Library of Congress did not agree to change the authority record to "Chinatown/International District," as according to LC rule interpretations there was not yet enough evidence to support the predominance of the new name. Despite that ruling, with the interest of the project partners and the end-users–most of whom will be King County residents–in mind, MIG approved the revision and suggested other groups catalog their images of that neighborhood using "Chinatown/International District."

HOW PURPOSE AFFECTS CHOICES OF DATA ELEMENTS AND VALUES

Libraries and museums have different missions and those missions were reflected in the approaches to metadata the different partners chose. Libraries, especially at research institutions such as the University of Washington, focus on providing facts that allow users to make connections, draw conclusions, and investigate further. The institution's job is to provide resources, not interpret them. Museums have a curatorial role to play and, as is commonly seen in online museum systems, most of the institutions involved in this project chose to extend that role to their cataloging. One partner museum included relevant essays containing contextual material in their metadata. Virtually all of them chose to include a *Caption* field, containing the type of information that might be found on an interpretive museum label.

The public's fascination with genealogy and the subsequently developed interest in local history mean that personal names are especially important to community museums. In almost every case the museums chose to emphasize proper names by separating out a personal name field from the general subject field. As the smaller museums are partly or all volunteer-run, and volunteers rarely have professional cataloging training, a metadata training workshop was organized for the participating project leads to help them avoid cataloging pitfalls. This is another area in which the University of Washington's Metadata Implementation Group was of great help. Their presentation at the workshop explained the importance of consistency, especially in the use of proper names. The standard vocabularies used in the King County Snapshots were new to most of the participants. Evaluation showed that by the end of the workshop 100% of attendees could complete seven of ten practice data entry fields correctly. Metadata specialists were able to resolve any lingering questions about geographic name standards and pluralizing subject terms, the main problem areas, during their on-site work. In this area, as elsewhere, the basic metadata entry performed by the metadata specialists provides examples for the museums as they enhance records in the future.

The metadata created walked a fine balance between conformity and reflecting each institution's style. Certain fields were required of all organizations for administrative and display purposes, such as the following:

Title

Digital Collection (the online collection name)

Ordering Information

Repository

Type (corresponding to *DC Type*–in this case, always image)

Digital Reproduction Information

Not only did these fields have to be present, but they had to have the identical field name.

Some non-required fields were chosen by all organizations, such as *Photographer*, *Date*, and *Subjects*. Other fields were used by all, or almost all of the museums, but not by the library: *Personal Names*, *Places*, *Physical Description*. For some of these, the UW Libraries chose an element name more compatible with standard library usage, whereas the field name chosen by the museums were more in line with museum usage. Instances of this include the museums' *Places* and *Physical Description* vs. the library's *Locations Depicted* and *Object Type*.

Image numbering systems proved to be an area where complete standardization was not possible, as the organizations had already established very different usages and wanted to track images in different manners. The following fields ended up as field names for the variety of numbers associated with images:

Image Number

Order Number (a number assigned at UW which is unique across its collections)

Negative Number

Image File Name (given by the scanner)

CD Number

Photographer's Reference Number

Serial Number

Some of the terms represent a slightly different type of number and consequently could not be combined, as the numbering systems provide the image (digital, print, or negative) identification and/or location information. Others were just variant names for the same idea. Many of these field names were not mapped to *DC Identifier*, as they are either for internal use only, or are references to the digital file, not the original photo to which the digital file refers. These fields can be seen in the software's administrative view which the museums use for internal searches and work.

Table 1 shows local field name mappings to Dublin Core elements. This shows which elements needed to be broken into more discrete information chunks, which ones had name disagreements, and which ones had the fewest variations.

TABLE 1. Dublin Core Mappings for All KC Snapshots Fields

DC ELEMENT	KC SNAPSHOTS FIELD NAME
DC Title	Title
DC Creator	Photographer, Architect
DC Subject	Subjects, Personal names, Classification, People/Places
DC Description	Caption, Notes, Comments, Additional information
DC Publisher	- no mapping -
DC Contributor	- no mapping -
DC Date	Date, Dates
DC Type	Physical description, Type, Object type
DC Format	Digital reproduction information
DC Identifier	Serial number, Image number, Negative number, Photographer's reference number
DC Source	Repository, Repository collection
DC Language	- no mapping -
DC Relation	- no mapping -
DC Coverage	Places, Location
DC Rights	Credit line
- no mapping -	Digital collection, Ordering information, Order number, Image file name, CD number, Administrative notes, Rig, Use, Classification

RESOURCE DISCOVERY

CONTENTdm's structure allows for maximum flexibility while retaining underlying structural connections via Dublin Core. Each collection is created as a separate database; consequently it can have any data fields that the creators choose to implement. Each field can be mapped to a Dublin Core element. CONTENTdm allows searching on an individual database, on any subset of databases on the server, or on all of the server's databases. The Dublin Core mapping allows multiple fields with semantic content equivalence within a single institution or in any number of institutions to be joined in cross-database searches regardless of field name.

At the time of writing, the native CONTENTdm search interface for cross-database searching, when not all fields have an exact name match, is to expose users to the qualified Dublin Core element set names. This is not very user-friendly. To get around this problem, and to provide a variety of means of access into the material for different uses, several alternative search mechanisms such as browse, keyword search across all fields, and pre-defined queries were added to the standard CONTENTdm Advanced Query mechanism in the KC Snapshots Web site. While each participating organization had detailed knowledge of their material, the project's staff and principal investigators had the best understanding of the body of material as a whole. Consequently, they were the ones who selected the pre-defined queries, such as *Military & War: Armories*, *Eating & Drinking: Bars & Taverns*, *Labor: All Labor Images*, and *Parades & Festivals: Parties*, which highlight areas of strength and topics of particular interest in the combined collections. Some of the queries used are fairly complex, and could probably not be successfully performed by users unfamiliar with the cataloging used in specific images. Each time a user selects one of these pre-defined queries, the selected query is run against all available data, so that even the most recently added material is returned.

A hidden searchable field called *Dates* was created and used by all organizations to allow searching on "circa dates." KC Snapshots is set up to allow the often imprecise date of the image to be displayed to the public in the *Date* field, which is not searchable, while allowing for a search on the associated *Dates* field. For instance, an entry in the *Date* field of "ca. 1965" would get an entry in the *Dates* field of "1960, 1961, 1962, 1963, 1964, 1965, 1966, 1967, 1968, 1969, 1970." Whether or not the small organizations will maintain this as they enhance the existing metadata remains to be seen.

The usage of *DC Subject* versus *DC Coverage* required some discussion. It was felt that it was necessary to separate the two for adherence to the Dublin Core Metadata Initiative. The *DC Subject* element is described in the Dublin Core element set as:

> A topic of the content of the resource. Typically, Subject will be expressed as keywords, key phrases or classification codes that describe a topic of the resource. Recommended best practice is to select a value from a controlled vocabulary or formal classification scheme.

DC Coverage includes the following text:

> Typically, Coverage will include spatial location (a place name or geographic coordinates), temporal period (a period label, date, or date range) or jurisdiction (such as a named administrative entity). Recommended best practice is to select a value from a controlled vocabulary (for example, the Thesaurus of Geographic Names

> [TGN]) and to use, where appropriate, named places or time periods in preference to numeric identifiers such as sets of coordinates or date ranges.

It was clear that the general public would not necessarily know or even guess that there was a distinction between these two fields. This dilemma was resolved with the realization that place names were geographic subdivisions applied to TGM I subjects, and would consequently appear in the *DC Subject* fields as well as in the *DC Coverage* fields. Local neighborhoods' names–an access point frequently requested by museum patrons–could not be put in *DC Coverage* fields, so they were entered in *DC Subject* fields. Names of built-works such as "Suzzallo Library" and "Chinook Junior High School" appear only in *DC Subject* fields. The Puget Maritime Historical Society chose not to have a field mapped to *DC Coverage*. This is because vessels–the subject of the majority of their images–are not place-bound, and were often photographed at sea where the geographic location is unknown. Even their images that feature people were often taken onboard ships with no identifiable landmarks to indicate where the ship was when the photograph was taken.

In the case of Puget Sound Maritime Historical Society's previously discussed fields *Rig* and *Use* and similar cases, a TGM term was used in the *DC Subject* field and additional fields which were not mapped to DC, but were indexed as local fields, were created for the more specific local terminology. The local fields can be easily searched from the organization's home page, where they are most relevant, but searching for them becomes more difficult on the unified Web site. In the unified site, they can be found using a keyword search, but an advanced search which displays Dublin Core element names requires the user to type the term in a field CONTENTdm labels "Across All Fields." This is because the field is not linked to Dublin Core and consequently does not show up on the default list of field names. "Across All Fields" is the software's equivalent of keyword search. However, TGM terminology can be searched without difficulty from both the project's and partners' Web pages.

BENEFITS FROM THE PROJECT

Cataloging of the twelve thousand local historical images was completed in the fall of 2003. When the grant ended, the project staff dispersed, leaving behind a group of committed and trained museum staff and volunteers to update and add images and metadata, with well-cataloged records as models, training materials as reference, and free use of Web-based software. It remains to be seen if cataloging consistency, and consequently users' ability to discover resources, will be ensured even without the cross-checking and staff support available during the grant-funded portion of the project. All of the participants have already enhanced existing metadata. Seven of the partners have stated that they would like to add new images into the database and a few of the partners have started discussing additional material that they would like to proceed with.

Having Dublin Core compliant metadata is a large step towards becoming an Open Archives Initiative Protocol for Metadata Harvesting (OAI-PMH) data provider. OAI-PMH participation would bring awareness of the KC Snapshots material and the project partners to a larger audience. CONTENTdm has the ability to export data in a variety of formats including tab-delimited, XML, and is also OAI-PMH compliant, so there is potential for becoming an OAI-PMH provider.

Collaboration on a project of this scope has led to increased contacts among the partner organizations, which serve as a support network and jumping-off point for future projects. It allows the partners greater understanding of each other's collection strengths. They can now more easily refer those interested in specific areas of King County history to the appropriate institution for their research needs.

All partners attended three workshops and an information session. At the information session, the benefit of talking with peer institutions was discovered. For example, those who already had some digital images online could allay the fears of image theft which concerned those who had not yet taken that step. In the workshop dealing with image sales, they were able to hear as a group the advantages of not undercutting one another in pricing and how to set prices according to the customer's pocketbook and intent. They can now view each other's metadata while examining and cataloging their own photos. This provides additional means of gaining information about their own materials, as many of the images held are of overlapping subjects and time periods. In these and other ways, increased communication will hopefully lead to improved museum processes.

ACRONYMS

AACR2–Anglo-American Cataloging Rules, Second Edition
DC–Dublin Core
IMLS–Institute for Museum and Library Services

KC Snapshots–King County Snapshots
LC–Library of Congress
MIG–University of Washington Libraries Metadata Implementation Group
MOHAI–Museum of History & Industry Seattle
OAI-PMH–Open Archives Initiative Protocol for Metadata Harvesting
PSMHS–Puget Sound Maritime Historical Society
SACO–Library of Congress Subject Authority Cooperative Program
TGM–Thesaurus of Graphic Materials
UW–University of Washington

REFERENCES

General Resources

Crossing Organizational Boundaries, King County Snapshots: A photographic heritage of Seattle and surrounding communities. [Accessed 28 October 2003] Available at URL: http://www.kcsnapshots.org.

DCMI, Dublin Core Metadata Element Set, Version 1.1: Reference description. [Accessed October 5, 2003] Available at URL: http://www.dublincore.org/documents/dces/.

DCMI, Dublin Core Metadata Initiative. [Accessed 5 October 2003] Available at URL: http://www.dublincore.org/.

DiMeMa Inc., CONTENTdm Digital Collection Management Software by DiMeMa, Inc. [Accessed 5 October 2003] Available at URL: http://contentdm.com.

Open Archives Initiative, Open Archives Initiative. [Accessed 13 October 2003] Available at URL: http://www.openarchives.org.

Pastime Software Company, Pastime Software: Museum collection management software tools & products. [Accessed 5 October 2003] Available at URL: http://www.museumsoftware.com.

University of Washington Libraries Metadata Implementation Group, Metadata Implementation Group. [Accessed 5 October 2003] Available at URL: http://www.lib.washington.edu/msd/mig/.

Project Participants

Black Heritage Society of Washington State, Inc., Black Heritage Society of Washington State: Preserving and exhibiting our past with our eyes on today and the future. [Accessed 14 January 2004] Available at URL: http://www.blackheritagewa.org.

Eastside Heritage Center, Welcome to the Eastside Heritage Center. [Accessed 14 January 2004] Available at URL: http://www.eastsideheritagecenter.org.

Maple Valley Historical Society, Maple Valley Historical Society: Preserving the past . . . Inspiring the future. [Accessed 14 January 2004] Available at URL: http://www.maplevalleyhistory.com.

Museum of History & Industry, MOHAI: Seattle's Museum of History and Industry. [Accessed 14 January 2004] Available at URL: http://www.seattlehistory.org.

Northwest Railway Museum, Northwest Railway Museum. [Accessed 14 January 2004] Available at URL: http://www.trainmuseum.org.

Puget Sound Maritime Historical Society, Puget Sound Maritime Historical Society: Preserving our northwest nautical heritage. [Accessed 14 January 2004] http://www.pugetmaritime.org.

Rainier Valley Historical Society, Welcome: Rainier Valley Historical Society. [Accessed 14 January 2004] Available at URL: http://www.rainiervalleyhistory.org.

Renton Historical Society, Renton Historical Society. [Accessed 14 January 2004] Available at URL: http://www.rentonhistory.org.

Shoreline Historical Museum, Shoreline Historical Museum. [Accessed 14 January 2004] Available at URL: http://www.shorelinehistoricalmuseum.org.

University of Washington Libraries Digital Initiatives Program, University of Washington Libraries Digital Collections. [Accessed 14 January 2004] Available at URL: http://content.lib.washington.edu.

White River Valley Museum, White River Valley Museum. [Accessed 14 January 2004] Available at URL: http://www.wrvmuseum.org.

Wing Luke Asian Museum, WLAM: The Wing Luke Asian Museum. [Accessed 14 January 2004] Available at URL: http://www.wingluke.org.

Vocabulary Resources

Getty Research Institute, Getty Thesaurus of Geographic Names Online. [Accessed 10 October 2003] Available at URL: http://www.getty.edu/research/conducting_research/vocabularies/tgn/.

Library of Congress Prints and Photographs Division, Thesaurus for Graphic Materials I: Subject terms. [Accessed 5 October 2003] Available at URL: http://www.loc.gov/rr/print/tgm1/.

Library of Congress Prints and Photographs Division, Thesaurus for Graphic Materials II: Genre and physical characteristic terms. [Accessed 5 October 2003] Available at URL: http://lcweb.loc.gov/rr/print/tgm2/.

Library of Congress, Library of Congress Authorities. [Accessed 5 October 2003] Available at URL: http://authorities.loc.gov.

Tacoma Public Library, Washington Place Names Database. [Accessed 10 October 2003] Available at URL: http://www.tpl.lib.wa.us/v2/nwroom/WaNames.htm.

Building a Common Catalog for Cultural Heritage Repositories: A Case Study of the Ohio Memory Online Scrapbook

Elizabeth Nelson
Laurie Gemmill

SUMMARY. The Ohio Memory Online Scrapbook is a common catalog with records from 328 cultural heritage repositories. Creating the online scrapbook, however, did not involve lengthy discussions of cataloging issues among the participants. Decisions about how materials would be described were made by staff at the Ohio Historical Society and were influenced by a commitment to integrating collections of different formats

---

Elizabeth Nelson, MSI, is Head of the Ohio Historical Society's Online Access Department of the Archives/Library Division, Ohio Historical Society, 1982 Velma Avenue, Columbus, OH 43211 (E-mail: enelson@ohiohistory.org). She also supervises the Ohio Memory Project and other efforts to make historical collections accessible via the Internet.

Laurie Gemmill, MLIS, MA, is Digital Project Specialist, OCLC, 6565 Frantz Road, Dublin, OH 43017-3395 (E-mail: gemmill@oclc.org). She was formerly the State Archivist at the Ohio Historical Society and supervised the Ohio Memory Project. She also serves as adjunct faculty at the School of Library and Information Sciences at Indiana University (IUPUI), where she teaches a course in digital libaries.

[Haworth co-indexing entry note]: "Building a Common Catalog for Cultural Heritage Repositories: A Case Study of the Ohio Memory Online Scrapbook." Nelson, Elizabeth, and Laurie Gemmill. Co-published simultaneously in *Journal of Internet Cataloging* (The Haworth Information Press, an imprint of The Haworth Press, Inc.) Vol. 7, No. 1, 2004, pp. 49-63; and: *Collaborative Access to Virtual Museum Collection Information: Seeing Through the Walls* (ed: Bernadette G. Callery) The Haworth Information Press, an imprint of The Haworth Press, Inc., 2004, pp. 49-63. Single or multiple copies of this article are available for a fee from The Haworth Document Delivery Service [1-800-HAWORTH, 9:00 a.m. - 5:00 p.m. (EST). E-mail address: docdelivery@haworthpress.com].

http://www.haworthpress.com/web/JIC

Digital Object Identifier: 10.1300/J141v07n01_04

from multiple repositories, a centralized approach, use of Dublin Core, development of metadata standards, and adoption of controlled vocabularies. Participating institutions, and end users, are more concerned with the results of the cataloging process than the details of description.

KEYWORDS. Ohio Memory, digital collections, primary sources, Ohio history, scrapbooks, digitization projects

More than 320 cultural heritage repositories in Ohio contributed records to a common electronic catalog without even knowing it. The participating institutions did not engage in endless debates about how to reconcile differences in descriptive practices or terminology. They did not retain a high-priced consultant or vendor. And they did not encounter any problems with legacy data. These institutions participated in the Ohio Memory Project, which was launched in 2000. The resulting product is the Ohio Memory Online Scrapbook, <http://www.ohiomemory.org>, a common catalog with more than 4,400 records and 26,000 images representing text and three-dimensional collections from archives, historical societies, museums, and other organizations that hold primary source materials and make them available to the public.

This paper will first provide background information on the Ohio Historical Society and the Ohio Memory Project and then explore five factors that contributed to the successful development of the Ohio Memory Online Scrapbook. They are: a vision of integration that eliminates distinctions between formats and physical locations, a centralized approach, the use of the Dublin Core metadata scheme, the development of metadata standards for the project, and the adoption of controlled vocabularies. Sharing some of the lessons learned through the project, this paper will be particularly useful to institutions planning similar efforts.

INSTITUTIONAL BACKGROUND

A non-profit organization incorporated in 1885, the Ohio Historical Society exists "to promote a knowledge of archaeology and history, especially in Ohio." The society, which is headquartered in Columbus, interprets, preserves, collects, and makes available evidence of the past. Moreover, it provides leadership on furthering knowledge, understanding, and appreciation of the prehistory and history of Ohio and of the broader cultural and natural environments of which Ohio is a part. Making collections available via the World Wide Web is a high priority for the society, even in a time of shrinking budgets. Digital projects are based in the society's Archives/Library.

There have been two phases of the Ohio Memory Project since its inception in 2000. Phase I ran from July 2000 to June 2002. The second phase concluded in October 2003. Although inital grant funding has ended, Ohio Memory will continue. Future plans include an education component, audio and visual digitization, and online exhibits. The mission of the project is to provide access to historical treasures of Ohio, bringing together primary sources from all parts of the state in an online scrapbook that commemorates the Buckeye State's bicentennial (which occurred in 2003); celebrates state and local history; encourages cooperation between archives, historical societies, libraries, museums, and other cultural organizations; and allows the global community to discover and explore Ohio's rich past. While the Ohio Historical Society is the lead institution in the project, the effort would not be possible without the support of six partner institutions: the Ohio Public Library Information Network, OhioLINK, the Ohio Library Council, INFOhio, the State Library of Ohio, and the Ohio Bicentennial Commission. These organizations provided financial, organizational, and technical assistance critical to the development, maintenance, and promotion of Ohio Memory.

The project is also indebted to the Ohio Historical Records Advisory Board, which originated the concept of a Bicentennial Scrapbook of historical materials from around the state. Although the name was changed to Ohio Memory, to capitalize on familiarity with the Library of Congress's American Memory, the project retained the scrapbook metaphor. It is fitting, because a scrapbook is a type of historical collection to be found in Ohio Memory, and because the term evokes the process of gathering together disparate materials that chronicle the history of a person or group and presenting them together with descriptive information.

Similar to an online public access catalog (OPAC) or a museum collection management system, Ohio Memory presents structured information about archival, library, and museum collections, but the online scrapbook differs in several significant ways. First, while there is a search engine similar to an OPAC, the results include images. Handwritten documents are also accompanied by text transcriptions. Linking to images reduces emphasis placed on descriptive metadata, because the records are no longer merely text surrogates for original materials; their value lies more in supplementing and contextualizing collections, providing background not just on the item but on the person, place, object, or event. The search process is not an interim step in resource discovery, but a direct connection to primary source materials and historical insight.

Second, the online scrapbook offers more access options than a typical online catalog. In addition to searching, users can browse materials listed by author/creator, contributing institution, county, format, place name, or subject. The county browse utilizes a clickable map of the state. Subject browsing draws on several controlled vocabulary lists, including a framework developed especially for the project.

Another important access method is My Scrapbook, a tool that allows users to create personal compilations of collections. While similar to a booklist in an OPAC environment or a shopping cart on a commercial Web site, My Scrapbook is more interactive. Users can select from several backgrounds and fonts to personalize the look of their pages. They can add their favorite collections (which appear as thumbnail images and item titles), create notes, and supply a title for the page. Each My Scrapbook has a unique URL, so that others can have read-only access. Notes enable users to supplement the general metadata provided by the project staff with information directly relevant to their interests and the context in which they are working. This makes My Scrapbook particularly useful in the classroom, since teachers and students not only discover Internet resources, but also collect and interpret them online rather than in a separate software program or on paper.

Project staff find the My Scrapbook capability invaluable, using it to highlight collections related to timely themes, such as Black History Month, New Additions, Ohio Presidents, Ohio's Bicentennial, Staff Favorites, Valentine's Day, and Women's History Month.[1] These featured scrapbooks encourage exploration of collections users might not otherwise find.

Although tools like My Scrapbook are becoming common for digital library sites, when developing Ohio Memory the project team found no models to copy and thus created it from scratch. Its appeal may relate to its similarity to online shopping. When a user adds an item to a shopping cart, there can be a sense of excitement about acquiring something new. Similarly, placing an item or collection in a personal scrapbook encourages Ohio Memory users to feel like the owner of a significant piece of history, at least in a virtual sense. Furthermore, by creating a note about an item, users are able to share their interests with others. Plus, they do not have to repeat searches for favorite items, which encourages repeat visits to the site.

A third way that Ohio Memory differs from single-institution catalogs is that it brings together online related materials in multiple formats that are held by geographically distant repositories. Items documenting the life of William McKinley, the 35th president of the United States, illustrate this well. There are more than 80 collections in Ohio Memory that pertain to McKinley, contributed by 13 institutions. Among these are text items, such as letters, newspapers, and sheet music; three-dimensional objects, including furniture, clothing, and campaign items; and audiovisual materials, mainly photographs. The materials trace his life, accomplishments, and legacy. While the majority of McKinley items are held by repositories in the northeast part of the state, where McKinley was born and lived as an adult, there are also items from western, central, and southern Ohio.

By visiting Ohio Memory, a user can form a more comprehensive understanding than they could by relying on just one institution's holdings and can view items without traveling across the state. A subject search for "McKinley and Civil War" displays his Civil War diary, kept while serving in the 23rd Ohio Volunteer Infantry, and the battle flag of the regiment, both from the Ohio Historical Society. These are complemented by a photograph of the young soldier in uniform from the McKinley Museum and National Memorial in Canton, and a special issue of the Toledo *Critic* newspaper published for an 1897 regimental reunion, held by the Rutherford B. Hayes Presidential Center in Fremont.

In the same way, the online scrapbook unites materials documenting McKinley's 1901 assassination. An image taken in Buffalo after the president's last speech and the mug shot of assassin Leon F. Czolgosz were contributed by the McKinley Memorial Library in Niles. From the Portage Historical Society there is a photograph of the funeral procession passing through Canton, located in neighboring Stark County. The Stark County Public Library provided a photograph of the interment ceremony.

VISION OF INTEGRATION

The McKinley items exemplify the vision of integration that underlies Ohio Memory. A strong influence on the project was the society's online collections catalog (OCC), which debuted in 1999. It provides access to MARC records that describe both print and three-dimensional collections held in the archives, library, and museum. Although museum collections are administered by a different unit of the society and are stored separately from archival and library materials, each collection type complements the other to provide more complete and compelling evidence of the state's history. Bringing together collections of multiple formats and from all collecting areas reflects the belief that it would benefit a patron interested in the Civil War, for instance, to not only retrieve records for military rosters, photographs, letters, and diaries from the war, but also military uniforms, weapons, and battle flags.

The educational and professional backgrounds of the original project staff also had an impact on the development of Ohio Memory. Half of the original project team had master's degrees in history, and the other half were trained as librarians and/or archivists. Although several staff had experience working with three-dimensional collections, professional practices of historians and information scientists were most familiar to the staff, which is reflected in the descriptive practices adopted for the project.

CENTRALIZATION

Having the core staff create all descriptive metadata for the project aids in the implementation of the vision of integration. The project was always envisioned as a centralized one, primarily because when the idea was first proposed in 1995, few organizations in the state had sufficient levels of digitization expertise and equipment ownership to make a decentralized project feasible. Since a fundamental goal was to involve as many institutions as possible, there was a definite need for one institution to take the lead, and to actively encourage participation by holding workshops and visiting institutions. Centralization, in the case of Ohio Memory, means that institutions submit descriptions and sample images or photocopies collections and staff select which ones will be included in the online scrapbook. The online scrapbook is intended to comprise high-quality historical materials, so staff apply rigorous selection criteria.[2] Nevertheless, the acceptance rate for the project exceeded 80%. Project staff worked closely with staff from contributing institutions to ensure that a high percentage of materials submitted would be accepted. Even if they met the criteria for historical significance, some collections were rejected because they had copyright or provenance issues, could not be digitized well, or duplicated materials already in the scrapbook.

Once accepted, staff arranged for digitization. Some institutions scanned or photographed their own materials, some items were digitized by another repository or vendor, and the remaining items were scanned by the project team at the Ohio Historical Society. Institutions that scanned their own materials or made arrangements for scanning by a third party were required to meet technical standards established for the project. As a final step before collections were added to the online scrapbook, staff created metadata descriptions based on information provided by the submitting institution and additional research.

DUBLIN CORE

Like the Ohio Historical Society's online catalog, Ohio Memory includes both print and three-dimensional collections related to archaeology, history, and natural history. Ohio Memory takes this idea of integration further by combining collections from multiple repositories. Unlike the OHS online catalog, however, Ohio Memory does not use MARC. Instead, it uses the Dublin Core metadata standard and XML, which are more flexible because the user can define which fields are used and how data is structured within each field. Images and metadata that comprise Ohio Memory are stored by OhioLINK, a consortium of Ohio's college and university libraries and the State Library of Ohio. OhioLINK's Digital Media Center, established in 1997, opted to make Dublin Core as its primary metadata standard because it is a "simple yet broad schema for indexing a variety of materials," allows for complex queries, and can be mapped to MARC or other formats.[3] For these same reasons, the project team agreed that it would be ideal for Ohio Memory.

American Memory, the digital library of the Library of Congress, was an inspiration for the project, yet not a direct model, since it does not include three-dimensional collections and was built around MARC and EAD rather than Dublin Core.

To design the online scrapbook, the project team essentially started from scratch. Ohio Memory was built, moreover, before the content was selected, digitized, and described. The team considered each element of Dublin Core and determined whether it would have value to users; in a parallel process, the team considered what information submitting institutions should be required to provide. The confluence of these lists, Dublin Core elements and data about collections, became the framework for a Microsoft Access database. Staff maintain the database at the Ohio Historical Center and export records to OhioLINK when complete.

Since the staff work primarily through Access and never see the records marked up in XML, the team is somewhat removed from Dublin Core. The situation is analogous to the Web markup environment. In the early days of the Internet, to prepare text for the World Wide Web required the use of a plain text editor and a thorough knowledge of HTML tags. Using a current version of an HTML editor requires only familiarity with word processing. And because Dublin Core fields can be defined by users, the project developed its own standards for the data that populate the Access database and change either the fields or the values as needed.

METADATA STANDARDS

Project guidelines specify the format of titles, descriptions, dates, place names, creators' names, and other details. Staff considered the needs of participants and users to be paramount, and tried to tie metadata standards to them. Information provided by institutions varied greatly, which can be attributed to differences in professional practices, provenance, format, and historical significance, among other factors. Staff evaluated the documentation provided by the participating repositories and attempted to bring uniformity to it.

Libraries and museums generally catalog or describe holdings at the item level. For example, there is a one-to-one relationship between a book and a MARC record or a painting and its catalog record. In contrast, multiple items that together form a manuscript collection or record series have only one catalog record and/or finding aid. This reflects the importance archivists place on provenance, the origin or source of a collection that provides the context needed to interpret it. Since Ohio Memory brings together library, museum, and archival collections, the online scrapbook interface must reconcile these differences in professional practice.

In resolving this issue, staff leaned closer to library and museum traditions. The majority of items are described individually. Some archival collections were even separated into multiple records. Size was one reason for this decision to deviate from standard archival practice. Since users navigate through large collections page by page, putting a large number of images under one record is less desirable because it reduces the chances that a user will see all significant pages. It is also problematic to describe a large collection of materials that include multiple subjects and topics in a way that the user can find the specific item they would find most helpful. Connections between related items are not lost, however, since there is a hyperlinked field for collection name, thus preserving intellectual relationships. Clicking on the name displays all items in a collection. Staff generally use the name provided by the institution, but in a few cases have created an artificial name when one does not exist.

For instance, there are, in the online scrapbook, five records that were unearthed from a time capsule placed in a cornerstone during an 1886 ceremony at the Mansfield Intermediate Penitentiary. Since it has closed, the prison has become popular with moviemakers–"The Shawshank Redemption," "Tango and Cash," and "Air Force One" were filmed there–and it is now a museum operated by the Mansfield Reformatory Preservation Society. The society has not cataloged the items, so had no collection name or catalog number. Ohio Memory staff assigned a collection title "Mansfield Reformatory Cornerstone Collection" to help users reassemble the items online.

While authors of books specify a title, creators of photographs and museum object typically do not assign titles. To standardize item titles, staff decided to present them with the first letter of each word capitalized. Initial articles such as "a" and "the" are omitted, as are titles like Dr., Captain, and President. If the item is a historical photograph, it is given a title that indicates both subject and format. An image of the Wright Brothers' milestone achievement is titled "First Powered Flight Photograph." [See Figure 1 for sample record and Figure 2 for the photograph itself.] Contemporary images of artifacts and buildings are given titles that indicate content only, so a photograph of the camera used to take the historical photograph (held by the Carillon Historical Park) is titled "Korona V View Camera and Tripod." The photographic format is simply a surrogate for three-dimensional objects, not a historical record, and thus is not emphasized in the title.

Other metadata elements–author/creator, date, description, subject, places, and contributor information–are more straightforward. If an institution has already cataloged an item or collection in MARC or another format, the submitting institution includes a copy of the record. Staff find this particularly helpful when creating metadata for Ohio Memory.

FIGURE 1

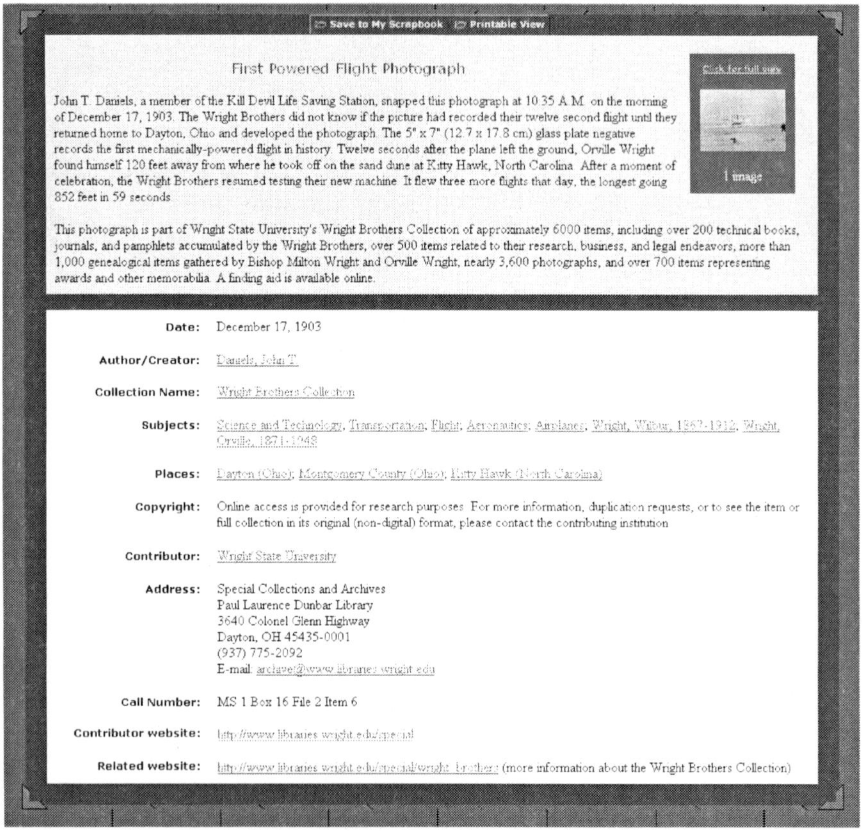

Reprinted with permission from Wright State University.

Prehistoric and natural history items are somewhat anomalous since they rarely have known creators. For geologic, plant, and animal specimens, such as the *Dunkleosteus terrelli*, or terrible fish, the Author/Creator field does not display online. For prehistoric collections, and any item of indeterminate source, the value in the Author/Creator field is "Unknown." When the information is available, the name is entered as in the Library of Congress Name Authority File or according to Anglo-American cataloging rules.

When an exact date is known, it is entered into the database in the format YYYYMMDD. In case of the Wright Brothers' image, appears as

FIGURE 2

Reprinted with permission from Wright State University.

December 17, 1903, although it is entered as 19031217 in the database. If the month is known, but not the day, it may be entered as YYYYMM, or 190312, which is displayed as December 1903. Dates may also be expressed as ranges, either with or without month and day. For approximate dates, circa is entered after the date so that it does not affect sort order. Dates of prehistoric items are expressed as 10,000 years before present.

Most descriptions are one to two paragraphs in length. The first paragraph presents the most basic information about the item or items: who, what, when, where, why, and how much (extent and dimensions). This last-named data element is one that staff have considered separating from the description field, but are unlikely to change because of the disproportionate amount of work entailed for a relatively minor benefit. While it is awkward to include measurements in a narrative description, a separate field might be confusing in some cases, such as when several photographs of different sizes are described in one record. Subsequent paragraphs deal with background or biographical information. Staff try to use as many synonyms as possible to facilitate keyword searching.

Some descriptions are much longer than others. There is a field limit of about 3,000 characters, so most descriptions have fewer characters. A few are longer and require special handling within the image management software. Many digital collections on the Web provide shorter captions, but the project staff felt that including more information would help users understand the historical records better. Since staff must research the collections to determine whether to accept them, it is worthwhile to include as much information as is available in the metadata record. The degree of historical significance influences the ease with which staff, both at the submitting institution and the Ohio Historical Society, can find supporting data by consulting available resources.

Place names are formatted according to Library of Congress Subject Headings. Contributor name, address, call number, and URL for institution Web site(s) are provided by the submitting institution, and edited for consistency in punctuation.

CONTROLLED VOCABULARIES

Use of controlled vocabularies is crucial to effective information retrieval. Ohio Memory uses Library of Congress Subject Headings, the Art and Architecture Thesaurus, and the Thesaurus for Graphic Materials. A framework of 21 locally-developed subject terms gives structure to the online scrapbook and facilitates browsing. Agriculture, Business and Labor, Science and Technology and Transportation deal with economics. African American Ohioans, American Indians in Ohio, Ohio Women, Immigration and Ethnic Heritage, and Religion in Ohio relate to the state's people. Civil Liberties, Military Ohio, Ohio Government, and Presidents and Politics pertain to governance. Architecture, Arts and Entertainment, Daily Life, Education, Literary Ohio, and Sports cover culture. Finally, Climate and Weather, Plants and Animals, and Geography and Natural Resources correspond with the natural world.

These locally-defined subjects have been altered slightly over the course of the project. A major change was the creation of a separate category for architecture, which was initially included with the other arts.

As the number of buildings in the online scrapbook increased, staff decided that it would be worth the effort to change the subject heading from Arts to Architecture, even though it had to be done record by record. Because it is such a laborious task, staff carefully consider such changes before making them. The software that manages Ohio Memory metadata (Documentum) does not easily allow global changes of subject terms the way that library systems do.

When assigning subjects, staff tended to emphasize simplicity and ease of use over rigid conformity, particularly to LCSH. For example, instead of "United States–History–Civil War, 1861-1865," staff apply the subject "Civil War" because it is shorter and because when looking through lists of subjects users are more likely to look for the war in the "C" section than the "U" section. Although few users notice, this deviation from LCSH counterbalances some of the benefits gained by using controlled vocabularies consistently. And while there are only a few people who assign subjects to collections, there is some variability in the application of subject headings that might impact on searching. It is to be expected, however, that if the project were decentralized and submitting institutions created their own metadata, there would be much greater unevenness.

In the sample record, one of the subject terms used is Flight. Clicking on the term in a record brings back more than 70 items. The earliest are childhood photographs and papers of aviation pioneers Orville and Wilbur Wright dating from the 1870s. Other items demonstrate the path of progress leading to Space Age items from the late 20th century, all connected by one subject term.

CONCLUSIONS

The Ohio Memory Online Scrapbook is a common catalog, but it is also a database, a Web site, an exhibit, a marketing tool, and a teaching resource. Which aspect of Ohio Memory predominates depends on the perspective from which it is viewed. Its collaborative and high-tech aspects garner the most attention. Although public libraries are well represented among participating institutions, few of the staff who were the primary contacts for their institutions claimed cataloging expertise. Even those who had a strong working knowledge of MARC recognized Ohio Memory as a new type of cataloging activity. The fact that the primary goal is not to catalog the items, but rather to make them more accessible, has helped the project to succeed.

When Ohio Memory was in the early development stage, there were few affordable off-the-shelf software packages for managing and providing access to digital collections. This forced the project team to develop its own ideas about how to proceed, and to build the online scrapbook interface around users' needs. Working in an institution that collects in multiple subject areas and formats helped the staff understand professional practices of several disciplines and adopt the most useful aspects for the project. Collaborating with OhioLINK, which uses Dublin Core, helped to define how the historical collections are described, stored, and displayed.

One of the forces driving participation in Ohio Memory was the desire for raising awareness of the institutions' holdings, particularly since many do not have Web sites or lack the resources to publish high-quality digital images on their sites. Organizations whose peer institutions had joined the project did not want to be left out. A small number of institutions had serious concerns about how their collections would be described and displayed, but most were willing or eager for project staff to assume that responsibility. And although contributing repositories are given the opportunity to comment on or suggest changes to the metadata applied to their records, most feedback relates to typographical and factual errors. In that way, the institutions are more like users, whose primary concern is the front end of the online scrapbook (content and display issues), not the back end (metadata and technical issues).

Based on their experiences with other Internet searches, users expect more than a MARC record. They want quick results, rich content, and something that they can use immediately, that does not require an in-person visit to a repository. Users are also accustomed to following hyperlinks to other sites to get more information. Metadata is important to help users understand the digital representations, although it is likely that some users simply skim the text. Using labels that make sense and displaying such information in an attractive way is important. Ohio Memory was designed with these considerations in mind. It has also been tested and refined since its initial launch, since technologies and standards change rapidly and sites that do not periodically refresh and expand are viewed as stale and uninviting.

Ohio Memory is a common catalog for 328 institutions, but it is also much more than that. It is something new that combines familiar aspects of cataloging with new methods of access and description. As a pioneer in this type of initiative, the Ohio Historical Society has had few models to follow, but it has been acknowledged, by the American Association for State and Local History and the State Library of Ohio and others, as an example for other institutions at both the state and local levels.

NOTES

1. For an example, see <http://worlddmc.ohiolink.edu/OMP/YourScrapbook?user=ohio200>.
2. See http://www.ohiomemory.org/om/participate2.html.
3. OhioLINK, *Standards for Multimedia Information: Humanities/Music/Archives*. n.d., (14 October 2003). <http://www.ohiolink.edu/media/dmcinfo/humanities_standard.pdf>.

SOURCES

Dublin Core. *Dublin Core Metadata Initiative.* <http://www.dublincore.org> (14 October 2003).

Library of Congress. *Library of Congress Authorities.* 4 November 2002. <http://authorities.loc.gov/> (14 October 2003).

Library of Congress. *Thesaurus for Graphic Materials I: Subject Terms (TGM I).* 1995. <http://www.loc.gov/rr/print/tgm1/> (14 October 2003).

Library of Congress. *Thesaurus for Graphic Materials II: Genre and Physical Characteristic Terms (TGM II).* 1995. <http://www.loc.gov/rr/print/tgm2/> (14 October 2003).

Ohio Historical Society. *Ohio Memory Online Scrapbook.* 14 October 2003. <http://www.ohiomemory.org> (14 October 2003).

OhioLINK. *Standards for Multimedia Information: Humanities/Music/Archives.* n.d. <http://www.ohiolink.edu/media/dmcinfo/humanities_standard.pdf > (14 October 2003).

Stories Told: Tribal Communities and the Development of Virtual Museums

Mark Christal
Loriene Roy
Antony Cherian

SUMMARY. Virtual museum projects that are partnerships between museums with Native American collections and tribally controlled schools offer a model of culturally responsive teaching practice and culturally responsible museum practice. In the projects students selected culturally relevant items from museum collections, imaged them in interactive digital formats, researched the items with the help of museum staff, academic professionals, and tribal elders and community members, and wrote essays to accompany the images in a virtual exhibition

---

Mark Christal, PhD, is Training Program Developer, Smithsonian National Museum of the American Indian, 4220 Silver Hill Road, Suitland, MD 20746-2863 (E-Mail: christalm@si.edu).

Loriene Roy, PhD, is Professor, School of Information, The University of Texas at Austin, 1 University Station D7000, SZB 564, Austin, TX 78712-3971 (E-mail: loriene@ischool.utexas.edu).

Antony Cherian is a doctoral student in Information Studies, University of Texas at Austin and Scholar-in-Residence at Varner-Hogg Plantation State Historic Site in West Columbia, TX (E-mail: cherian@ischool.utexas.edu).

[Haworth co-indexing entry note]: "Stories Told: Tribal Communities and the Development of Virtual Museums." Christal, Mark, Loriene Roy, and Antony Cherian. Co-published simultaneously in *Journal of Internet Cataloging* (The Haworth Information Press, an imprint of The Haworth Press, Inc.) Vol. 7, No. 1, 2004, pp. 65-88; and: *Collaborative Access to Virtual Museum Collection Information: Seeing Through the Walls* (ed: Bernadette G. Callery) The Haworth Information Press, an imprint of The Haworth Press, Inc., 2004, pp. 65-88. Single or multiple copies of this article are available for a fee from The Haworth Document Delivery Service [1-800-HAWORTH, 9:00 a.m. - 5:00 p.m. (EST). E-mail address: docdelivery@haworthpress.com].

http://www.haworthpress.com/web/JIC

Digital Object Identifier: 10.1300/J141v07n01_05

on the World Wide Web. Students learned about museums, new imaging technologies, and new ways to communicate their culture to the world. Tribal members saw the projects as mechanisms for improving relations between Native communities, museums, and non-Native communities. Museums responsibly served their Native constituents and gained new community-based information about their collections.

KEYWORDS. Virtual museums, cultural heritage programs, American Indian communities, culturally responsive education, culturally sensitive collections, school-museum partnerships

Since 1999, Native children, their educators, and Indian community members have been collaborating with museums such as the National Museum of the American Indian (NMAI) to digitize cultural objects held as part of the museums' collections. Most notably, these efforts resulted in the development of a virtual tour of two exhibits on display at the NMAI's George Gustav Heye Center in Manhattan. We have reported on the technology we used and on the process we followed in several articles, a dissertation, and at conference presentations including the annual conference of the American Library Association, the first International Indigenous Librarians Forum, the Barents Libraries Conference, ED-Media, the National Oral History Association Conference, the National Association of Native American Studies, and the Museums and the Web 2001 Conference.[1] In short, children attending tribal schools traveled to the George Gustav Heye Center in Manhattan to create three-dimensional movies of objects that were de-installed from exhibits and moved temporarily to an impromptu photography lab. These object movies were linked to panoramic movies of exhibit areas, creating a virtual space that mimicked the experience of being in the museum space. The students also wrote essays interpreting the objects. These essays became the HTML text labels for the virtual museum. The resulting Web site at http://www.conexus.si.edu/vrtour/ includes the history of the project, image maps of the floor plans of the two museum exhibits, text labels, thumbnail photographs of the digitized objects, and the linked virtual tour. With the free QuickTime plug-in from Apple Computer, Web site visitors can view the virtual tour in any standard Web browser.

In this article we will review these efforts from another perspective: how digital images of tribal cultural material help to tell the story of contemporary indigenous communities. We will also discuss issues concerning sharing indigenous cultural heritage in a publicly accessible online environment.

REPLICATING THE MODEL

The school-museum partnerships that produced the virtual tour of the Heye Center exhibits in 1999-2000 provided a general model that has been replicated several times at the NMAI and in other museums in recent years. The projects, or "digital workshops," vary according to the goals of the participants, but the common elements of these projects provide a general model. Native students select and image items from the museum's collections and research those items with the help of their teachers and tribal community members. The students then write essays or "text labels" of the items for the virtual exhibit. The model is simple and flexible. It achieves a major goal of the museum's mission: to make Native collections more accessible to Native people.

Christal applied the model to a number of school-museum partnerships that matched tribally controlled grade schools with regional museums that housed relevant Native American collections. Four of those projects became case studies for his dissertation, which examined virtual museum projects as a culturally responsive teaching strategy.

The National Museum of the American Indian has adopted the model in a series of digital workshops with tribal schools and colleges that have been taking place at the NMAI Cultural Resources Center in Suitland, Maryland, where the bulk of the museum's collections are stored. In the summer of 2001, the museum brought a group of middle school students, teachers, and community members from Little Wound School on the Pine Ridge Reservation in South Dakota to photograph winter counts for the cultural education program. Winter counts were created by several Plains Indian tribes as a way of recounting the history of the people. Each winter, the elders would convene to decide what event to commemorate that year. The keeper of the winter count would paint a picture or "glyph" on a tanned buffalo robe to represent the event. On occasion, the winter count would be presented to the people to help them recall their history.[2] For the Little Wound workshop, the students digitally imaged two winter counts, along with several other items from the collection that further illustrated the history and lifeways of the people. One winter count represented 105 years of the history of the Hunkpapa band of the Lakota from 1798 to 1902. The other, known as Lone-Dog's winter count, covers seventy years of Yanktonais Dakota history, from 1801 to 1871. The students learned how to use digital cameras to photograph close-ups of each glyph on the winter counts and how to use Adobe Photoshop to assemble the pictures into a linear timeline. The Little Wound culture teacher and a tribal elder recited audio recordings that interpreted each glyph in both Lakota and English. The students wrote short essays about some of the historical events depicted in the winter count. For instance, several glyphs on both winter counts memorialized the effect of smallpox and other diseases upon the tribes. One student researched the history of European diseases introduced into Native America and wrote an essay that included links to those years on the winter count. In the Web version of the winter count, clicking on any glyph in the timeline will bring up a window showing a close-up of the glyph and the associated audio files, and links to student essays that relate to the event. The student essays also link back to relevant years of the winter count. Social studies teachers who have seen the winter count material have been very enthusiastic about it because they have such a difficult time finding materials on 19th century American history created from the perspective of Native people.

In the summer of 2002, the NMAI conducted two digital workshops with middle school students from Ahfachkee Tribal School (Big Cypress Seminole, Florida) and museum studies students from Haskell Indian Nations University in Lawrence, Kansas. The workshops were conducted in partnership with the Weedon Island Interpretive Center of Pinellas County, Florida. Ahfachkee students made three-dimensional images of artifacts from ancient mount-building cultures, their ancestors from over 1,000 years ago. The media and text they produced was then developed for interactive touch-screen information kiosks at the interpretive center and a virtual tour Web site (http://www.weedonislandcenter.org). The Haskell students not only helped the Ahfachkee students learn the virtual museum technologies and assisted them in their research, but they also selected items from the collections to image for an exhibit at the Haskell Tribal Museum and Cultural Center.

Two additional digital workshops were conducted in 2003 with communities that will be represented in the upcoming exhibitions in the new National Museum of the American Indian on the National Mall, that opened in September, 2004. The materials the participants produced in those workshops were developed for a Web-accessible virtual tour to help celebrate the new museum opening. The two groups were from the Anishinaabe community of Fond du Lac (Minnesota) and the Yup'ik community of Alaska.

IMPACT ON STUDENTS

Responses from the student participants in these projects open one window to the cultural stories emerging from our school-museum partnerships. Students in the NMAI virtual museum project kept diaries during their visits to Manhattan in 1999. These diaries detailed their experiences in the City:

> What did I like about New York City? There was a lot of cheap stuff you could buy in Chinatown, for example, yoyos, t-shirts, and watches. I liked the big buildings cause they were so tall that it could make you dizzy.

> What I liked in New York was the subway. The first time we rode it all of the teachers flew back and hit each other . . . I liked the mike [*sic*] and pizza cause they were made by the real Italians.

The children reflected on what they learned about the NMAI:

> What I learned in the museum was . . . the objects could only be handled by museum personnel. They had to handle the objects very carefully with gloves.

And they described how they contributed to the digitizing projects. They outlined the new technology skills they had acquired:

> What I learned about making panoramas we had to make sure you didn't have any reflections in any glass exhibit boxes.

> I also learned that you have to turn the turntable really slow so the objects won't fall off the turntable.

When the initial project at the Heye museum took place, only the items on exhibit were available for the student choices. The Smithsonian Institution had acquired the collection from the Heye Foundation in 1990, and the nearly one million items stored in the Bronx facility were in the process of being moved to the Cultural Resources Center, the new collections facility in Suitland, Maryland. This situation meant that the students had little or no access to items that reflected their specific cultures. These limitations led us to conceptualize the project as a virtual tour of the current exhibitions as seen through the eyes of Native children. By digitizing cultural material from other indigenous communities, the children compared these cultures with their own. In their diaries they spoke about what they learned about other indigenous communities–those who were represented by the objects on display and by the indigenous museum staff who assisted them in their work.

> I learned, that Mimbres people put a hole in their bowls to release their spirits.
>
> Then I learned that the Inka people used Jaguar Goblets for sacred ceremonies. The second day I learned that people use the Gourd Rattles for sacred dances like an animal dance. I also learned Zapotec people use Urns for sacred ceromonies. Also kings use them.
>
> I liked Juanita's (a Mayan museum docent) cooking of the stew, tamales, and the tamale sauce.

The children followed a structured approach in investigating the creation and use of the objects they digitized. They used library resources and interviewed curators and other museum personnel to answer these questions:

- Who created this artifact?
- Where did these people live?
- What was the object's purpose? Why was it created?
- How was it created?
- Why did I select this object for the tour?
- How is this object connected or not connected to my culture?

The text labels they created not only answer these questions but also connect a contemporary Native person to times past (that is, to the creator of the object and his/her community) and the future (to those who will discover and interact with the virtual tour).

There were a few cases when the students were able to make selections that came from their own communities. For instance, one student discovered a pottery plate on exhibit that had been made by his great aunt. His text label for the virtual exhibit revealed his pride and cultural knowledge and foreshadowed the kind of intercultural communication that may result from such a close cultural matching.

BLACK-ON-BLACK PLATE BY MARIA MARTINEZ

> My aunt Maria Martinez made this pottery plate. She is famous for her pottery work. When I was coming to New York, there was a pot that she made in the airport. When she gathered the clay it was dirt. She sifted the dirt to remove stones. Then she wet the clay and made a piece of art. She polished the surface with round, river rocks to make the pottery shiny. Then she brought the pottery outside, covered it with dry cow manure, and put tin and metal sheets on top of the wood she used to fire the pottery. Some black-on-black pottery has designs of kiva steps, rain clouds and feathers.
>
> *–Galvin, Santa Clara Day School*

Once the children and their teachers created the digital images, they returned to their tribal communities to discuss their experiences: "on returning home, they consulted with tribal elders about the appropriateness of their choices and the unique histories of the objects."[3]

For the remaining virtual museum projects and digital workshops, there were fewer restrictions on what items in the museums' collections the students could choose from. Close cultural matching between the student and the artifacts became the rule rather than the exception.

The Yup'ik digital workshop in July 2003 included both high school and college students from southwest Alaska. These students responded to the project with great interest and enthusiasm. As in all the projects conducted at the museum, cultural connections were prime motivators for the students. One high school student, Elsie Alexie talked about seeing items she had only heard about before:

> My grandmother used to talk about some of the things I photographed . . . I got to see some of the items she used, like a clay pot-

> tery lamp. When I told my grandmother about it she was so proud that I got to actually see these things.[4]

College student Dawson Hoover clearly valued the opportunity to see rare items such as the rain parka made of fish skin. He said, "I had heard about this in high school, but to actually see one was a totally different thing. A lot of my peers will never get to see one." Like many Native people, young and old alike, he felt a deep spiritual connection to his ancestors through the objects. He explained:

> I could feel the energy coming from some of the items, but this particular mask was the most powerful. That was something I will never forget. I wish we had these items to see here.[5]

These statements from two young Yup'ik participants illustrate the qualitatively different resonance that Native people find in museum items that come from their own specific cultures.

The virtual museum projects impact more than those involved with their construction but involve the wider tribal community and those who visit and interact with the virtual tours available in local museums and to the world through the Internet. The virtual museums can also help instill a sense of local pride. According to Marty Kreipe de Montano, site manager for work conducted at the National Museum of the American Indian, "Our grandparents were told to forget about our traditions. These kids will grow up with the knowledge that our cultural traditions are worthy of a great museum."[6]

CULTURALLY SENSITIVE COLLECTIONS

The National Museum of the American Indian Act and Amendment of 1989 provided for the establishment of the Smithsonian NMAI.[7] The Native American Graves Protection and Repatriation Act (NAGPRA) was signed into law in 1990.[8] Both of these acts call for the repatriation of human remains and funerary objects, sacred objects, and objects of cultural patrimony to the Indian tribes or Native Hawaiian organization that can show that the objects were once owned or controlled by the tribe or organization. Both of these acts (the first specific to the Smithsonian, the second to other museums that accept federal funding) required that museums cooperate with Native tribes to identify and inventory all objects in museum collections subject to repatriation and to respond to repatriation requests.

Initially, many museums were concerned that repatriation requirements would decimate their collections, but that has proved not to be the case.[9] The NMAI has been especially proactive in its repatriation program and has to date repatriated about 2,000 objects, a tiny proportion of its collection of over 880,000 objects.[10] Instead, by cooperating with Native people and opening up their collections in the consultation process, museums have gained new knowledge about those collections that comes directly from the Native communities, a new understanding of the significance their collections have for Native communities, and a renewed commitment to serve their Native constituents.

Since its founding, the NMAI has implemented a policy that integrates Native worldviews in all aspects of museum activities. The museum consults with Native people to determine the appropriate policies and protocols for documentation, acquisition, repatriation, curation, exhibition, and other museum activities. Because the concept of the sacred is central to the worldview of indigenous cultures, the museum also consults with Native people to determine and implement traditional care and handling of museum objects that is in keeping with Native spiritual sensibilities.[11] In some respects the notion of "culturally sensitive objects," a term that describes objects subject to repatriation, may also be applied in varying degrees to all the objects in the museum's collection. To honor and respect the spiritual qualities of the collections, the NMAI brings Native spiritual leaders to its facilities every season to offer prayers or ceremonies recognizing the sacred aspects of the collections and to bless the museum's activities. All museum personnel, Native and non-Native alike, are invited to participate in the seasonal blessing, and many do. The Cultural Resources Center, where the bulk of the collections are stored, has a special room where prayer and ceremonies may be performed in private. There is also a fire circle in a nearby grove of trees where dancing, singing, and ceremonies may be performed.

The openness to indigenous worldviews has affected the approach to the virtual museum projects, especially with respect to objects that may be culturally sensitive. For instance, when the Seminole students came to the museum to image objects from the ancient mound builder cultures to which their distant ancestors belonged, the project's cultural advisors asked to bless the facility before the project began. The NMAI seasonal blessing coincided with the first day of the project, which satisfied the advisors' request. The Anishinaabe group wished to purify the items they were imaging at the beginning of each day, so every morning a cart bearing the objects for the day's activities was brought into the prayer room for smudging. All of the NMAI staff involved in the project were invited to participate in the ceremony.

The first virtual tour project often involved Native children working with items not from their specific culture. For this reason, some items the students selected had to be substituted with another object because of culturally sensitive issues. For instance, one student selected a headdress for imaging that was embroiled in controversy. The museum had consulted with the tribe in good faith, and tribal representatives had approved the headdress for the exhibition. However, a newly elected tribal government objected to the exhibit. After consulting with a tribal member, the museum decided that the new situation required that the item not be included in the virtual tour, and the student selected a different headdress to image. This example demonstrates the need to treat the use of museum objects on a case-by-case basis. Approval to use an item in one context does not necessarily mean is it approved for all contexts.

This culturally sensitive approach to materials is a natural consequence of consultations with the Native communities represented by the museum's collections. Those who work with tribal communities must not only have a respect for Native knowledge but also must accept and support restrictions on access and use of tribal information. In 2000, the Assembly of Alaska Native Educators adopted "Guidelines for Respecting Cultural Knowledge" with protocol statements for Native elders, authors and illustrators, curriculum developers and administrators, educators, editors and publishers, document reviewers, researchers, Native language specialists, Native community organizations, and the general public.[12] If technology provides an opportunity for easy access to Native culture, it can also be used to misrepresent tribal culture. However, museum specialists involved in indigenous knowledge imaging are also beginning to explore use of technology to provide Native communities with control over virtual museum text labels and imaging.[13]

THE IMPORTANCE OF ACCESS

Christal completed a dissertation study of four virtual museum projects by tribal schools at regional and tribal museums that documented responses similar the ones observed at the NMAI.[14] His analysis addressed two questions: first, the process of developing the virtual museums and, second, how the virtual museum efforts impacted teaching that was culturally based. His study revealed a spectrum of responses that students, teachers, and Native community members had to virtual museum projects. A number of themes in the study related to the unique access that students and community members gained to important museum objects related to their cultures.

One theme that emerged from the study was "bringing objects out from behind the glass." This metaphor treated the glass of museum display cases as symbolic of the limited access Native people have to the important cultural holdings of many museums. Anna, the cultural program director at Pacific Tribal School first introduced an image of "bumping your head against the glass"[15] to get a better view of museum exhibits.[16] Anna also referred to her initial reaction to the objects on arriving to the museum to begin the project. She wanted to give thanks for the opportunity at that moment, because she realized "this was someone's at some time. Somebody actually made this, and a trace of time is carried within them. I almost felt like crying."[17]

Other Native participants in the school-museum projects in Christal's study talked about sensing a connection to their ancestral past through the physical presence of the museum items. Wendy, a liaison to the Wabanaki people (northeast U.S. and southeast Canada) for a Canadian Archeological Commission also spoke of the connection that was made between the Native participants and the artifacts. She said, "there was an interest there, a great interest here, and again, I feel it is because of the energies, the energies from the artifacts, from the items."[18]

Frances, a member of the museum planning board of the Pacific tribe expressed the significance of the access provided by the project.

> I think the thing I'm most excited about is that we're allowing the kids to have ownership of those objects. "These are yours, you tell us what they mean, you tell us what they were meant for, and you tell us what they mean to you as an individual." Being able to let the kids have that voice now is going to give them a sense of who they are, a deeper sense than if they were being just talked to by the curator about what those things are interpreted as. But being able to, as a tribal member, to sit and listen to my aunt talk about her experience in life, and being able to listen to her share with the curator, and for him to appreciate it and take her word and everybody comes away with a lot more insight and a lot more excitement about the pieces, that they're not just a part of someone else's history that's long gone. They're part of a history of a people that's still thriving. The people that used those still use them to this day, and I think that's what–it kind of pulls them into the here and now more than leaving them in the dust and the dark, behind

> the glass . . . I don't want my culture to be viewed only from behind the glass.[19]

Here, Frances is referring not only to the access to the objects the project afforded the students but also to the effect of the project design as a whole. The students were given the opportunity to select the items that spoke most to them. Students also got to see how museum staff respected and valued not only the museum items, but also the knowledge of the tribal elders who participated in the project, providing a validation of their culture they rarely encounter in mainstream society. Frances' main point is the connection to the students' heritage that was made through the close encounter with the museum objects. As they were putting one of the items back into the museum case, one student whispered to her, "Just think, we'll never be this close and have this kind of access ever again."[20]

IMPROVING RELATIONS WITH THE NON-NATIVE COMMUNITY

Frances' remarks also touch on another strong theme that developed in Christal's study: the role of the virtual museum project in improving relationships between Native and non-Native communities. Participants in all of the projects remarked how putting the virtual museum materials–images and student writing–on the Web would work to counter the stereotypes and misconceptions about Native people that have grown out of prejudice and misrepresentation in mass media. For instance Annie, a Wabanaki member of a public museum board of directors complained that

> The culture of the Plains and the Southwest gets imposed on us. The culture of this area is really not well known and not associated with what's traditionally Indian . . . I think that any kind of education, public education, about Indian people serves to get rid of some of the stereotypes that are out there.[21]

She went on to explain that many people were not aware of the American Indian people in their midst. She said, "It [the virtual museum] puts us on the map, first of all, and people can say, 'Oh, yes, there are Native people in the northeast,' which is something."[22]

Catherine, an Anishinaabe technician at the Great Bear Tribal Museum, hoped the virtual museum would be a marketing tool that would bring people to the reservation and the tribal museum so they could experience contemporary Native American life first-hand. She thought the virtual museum provided a "phenomenal opportunity to defictionalize" Native Americans. Catherine also made the complaint that has often been lodged in Indian Country, that many people in mainstream society still believe Native Americans live in nineteenth century conditions. She believed that

> this museum will show where the Great Bear people are from, how we got here, what's happened since then, and our current comparatively affluent [lifestyle] . . . We're modern people. We're modern people with a very deep connection to the past.[23]

Fred, an Anishinaabe basket maker and tribal historian saw the virtual museum project as a way for Native people to assert themselves in a positive fashion. He said, "I think it helps the community to say who they are and what they have to offer the world, helps us to clear up misconceptions, . . . and it give us the opportunity to share to the world."[24]

Frances was excited about the prospect of the virtual museums being used as a part of the public school curriculum that has been lacking quality educational material about Native peoples.

> With these kinds of projects the access will be in an arena, maybe middle school, elementary school kids, and so when you are able to reach the kids through technology that they are so used to–it's just everyday life to them now to open the computer Web page and browse the Internet–that the access is a little more freer of judgment, and you're not looking at the textbook that was written in 1950, interpreting life for Native Americans from that point of view, but this is an interpretation by the kids, by the Indian kids themselves, and so the people who access that information are going to get a more accurate picture and a more contemporary picture of Indian life, past and present.[25]

STUDENTS AS CULTURE BEARERS

Frances' statement reflects the belief shared by other community member participants who thought that it was especially important that it was Native young people who where creating the virtual museum materials and writing the interpretive essays. Will, the cultural program director at White Pine Tribal School (Wabanaki), expressed it this way:

> There's a lot of people that know–what they know about us is not true . . . If we can present ourselves in a way that we want to, and we have control over that, it makes it so much easier and less tense because lots of times we are misrepresented. I mean, even in school, where you have mascots that depict us in a not very complimentary way. This is another way for our own students to talk to those students: "[That] is not the way we really are, this is what we really are, and this is who we want to be depicted as," you know. And I think at that point, more serious intercultural learning takes place. Sometimes when it's at the political level it doesn't, not much learning takes place there, but mostly negotiating, but at the kid-to-kid level, there are some genuine exchanges there. It works better than most people give it credit for.[26]

By participating in the "kid-to-kid" intercultural communication expressed through the virtual museum, young tribal students were learning to take on the very important role of culture bearers for their community. Community members expressed this in various ways. The students were becoming the "storytellers for their grandparents" or "representatives of the future" for the tribe.[27]

ADVANCING TRIBAL MUSEUM EFFORTS

Tribal museums and cultural centers are becoming a common feature on tribal reservations and in Indian country. Native people who work to promote these institutions have shown great interest in virtual museum projects as a way to advance their goals and enhance their existing programs. When Marty de Montaño of the National Museum of the American Indian showed the Heye Center virtual tour to members of a Tohono O'odham culture committee, they believed it would help persuade some of their own community members who were resisting a tribal museum plan because they believed that museums were about displaying the bones of the ancestors. Frances talked about the problems that the Pacific Tribe was having getting the funding for a tribal museum. She thought the virtual museum would be a "jumpstart" for that effort.

For the tribes that had established museums, the virtual museum projects were a way of promoting their institutions to a wider audience. The projects also suggested ongoing programs that might be established in partnership with the tribal schools. Rose, the Gifted and Talented Program teacher at Great Bear Tribal School, wanted to establish a program in which students would do imaging and writing about items as they were donated to the tribal museum. The close proximity of the museum to the new school (about 100 yards) made such a project feasible. Northern Lakes Tribal School (Anishinaabe) also did not have a tribal museum, but their experience in several virtual museum projects inspired a proposal for a museum that would be run by the school for the purpose of imaging items from the community and doing oral history projects.

The general notion of forming digital collections of objects within the community occurred to several project participants. Frances shared some concern about preserving cultural items within the community:

> I think that a lot of community members might have objects themselves that they would like to preserve in a 3D setting, so that their kids–in case something should happen and someone doesn't value it as much as they do–that it would be on record somewhere . . . Yeah, I think that's really important, and then the elders want a place to be able to put this information in writing or in record.[28]

The Pacific Tribal School technology coordinator realized that museum-like conditions are difficult to replicate in someone's home. Doing digital collections within the community might provide one way of recording family heirlooms for the benefit of the tribe, because

> when you see how careful the museums are just to keep it almost in no air, no light, no everything, and who can do that? So I think people really would be happy to have [their keepsakes] digitized so that they can share them without sharing them.[29]

THE NATIVE RESPONSE TO TECHNOLOGY

Virtual museum projects may prove to be a tool for cultural reproduction for Native people, but it is a departure from the traditional forms of intergenerational learning. Native people have shown some healthy skepticism regarding the promise that Western technology might hold for cultural survival. Technology has proven to be a double-edged sword with the potential of doing harm as well as good. According to Native scholar Delgado, many Native people mistrust modern technology because it is "a product of the corporate western mind; we [Native people] do not know that the sacred was part of the thought process that produced these objects. For some Native people, this concept–being surrounded by the non-sacred–goes against the very tenets of Native life and thought."[30]

Christal asked the Native participants in his case studies if they thought that the use of technology was an acceptable tool for cultural reproduction. Several participants indicated that there was a generational shift toward more acceptance for technology use, but there was still great concern about the potential for abuse. Native people have demonstrated their ability for adopting technologies in the past that served as tools for survival. Will, the culture program director at White Pine Tribal School, indicated the need for caution in adopting Western technology in the context of being in a cultural "survival mode."

> I think that I'm a little bit nervous when we put all of our stories out there on the World Wide Web. I think you don't put out all of them because, and this is from the history of manipulation, you have to be really careful. You want to teach the outside world, but you have to be careful about how you do that, and until there's a better sense of trust. And some of the elders do not trust that much because all they see is another outside entity trying to destroy us. That's a real thing. So we kind of have to discuss that and talk about that. But I think that at this point we have to spend so much time at the survival part. If we don't do something in the next forty years or fifty years we might not have a language working. And so whatever we're doing right now is meeting those purposes. And we work hard at it. It's difficult.[31]

When asked if he thought technology was part of the problem or part of the solution to language and culture loss, Will replied, "I think it is part of the solution." Will is expressing a need for balance between adaptation to modern technology and survival that was expressed by other participants in the virtual museum projects. Jack, the Anishinaabe technology coordinator for the Great Bear Tribal School explained,

> We've got to adapt to it. Everything has been handed down from generation to generation, but now we're to the point of where the elders are getting very old, and our younger people who are sup-

> posed to pass it down have no interest in it any more. I see technology as a way of preserving it. They tell their stories to us, which will be recorded, documented, again, ask the elders if it is shareable or not with the general public, but there's a whole lot on the reservation that can all be documented. It could be stored here and shared with the culture on the reservation, but then with the elders' advice they could share what they wanted with the outside world.[32]

For Jack, consultation with the tribal elders was a key component in the virtual museum projects, because they advised on which aspects of culture may be shared with the outside world in digital form and what aspects should be kept within the community.

Other participants however, saw a greater urgency to preserving what remained of culture and expressed less need for caution in adopting current technology. Ernest, the curator of a Wabanaki tribal museum explained that

> We're some of the first Native people who were encountered [by European colonists], and [we] have been surrounded by a non-Native culture for so long and we've lost so much that I can see that's what we have to do. We have to grab what we can and put it down the fastest way in the hardest copy that we can and try to keep our traditions, if only that way, because, as these kids grow up, they're growing further away from the culture . . . But if we can get these stories and parts of our culture taken down to be shown, get a new museum, they'll be more community involvement there. More people will see it. There will be more of a tide that will grow out of this. It's like a seed that's been planted. It's the best of two worlds. It's our culture on one side and new technology on the other side, and with these two things coming together, it makes the seed of our stories in the future, and for our future generations of people.[33]

Wendy was concerned that the survival mode that the Wabanaki people were in might lead them away from an appreciation of their own culture. "They want to succeed," she said, "and so they go along with whatever is coming up or whatever is the hot occupation or whatever, and technology is." She saw in the virtual museum project a way of putting technology and culture together, so that "you don't have to make a choice between culture and technology . . . No, you can have them both."[34]

Fred saw no contradiction between modern technology and Native American traditions, "because it's a tool, and it's a gift from the Creator, like anything else, and it's for everyone's benefit." Fred made a more explicit connection to the virtual museum activities and commerce between Native people that has been going on for millennia.

> I really think that it's the tribes getting involved with each other and them [getting] to share cultures among the tribes themselves, not only the culture, but the technology and the different things that go along with it. And that's important too–it's sharing, always that sharing. Yeah, I think it's like when we used to always trade like beads for shells and things like that, and now it's sharing the technology.[35]

CORRECTING AND ENHANCING THE MUSEUM KNOWLEDGE BASE

The findings from Christal's study discussed above show a variety of responses to virtual museum projects by Native participants. As in any true collaboration, all participants benefit, including the museums. The experience at the National Museum of the American Indian has shown that the virtual museum projects have been especially beneficial in terms of correcting and enhancing the information on its collections. The bulk of the collections for the NMAI were collected by George Gustav Heye, who became a passionate collector of Native American materials after buying his first item, a deer skin shirt, from an Apache woman in 1897. His hobby soon became an obsession. After gaining his fortune in investment-banking he was able to collect Native artifacts by the boxcar load.[36] He sent agents and organized expeditions over the whole Western hemisphere, purchasing everything he could–from tribal people, dealers, and estates of other collectors. In 45 years of collecting he had amassed the collection that was to become the basis of the Smithsonian National Museum of the American Indian.

Heye kept record of each object on 3″ × 5″ index cards. A majority of those cards indicate little more than the name of the item, which tribe it was associated with, where it was collected, and by whom it was collected. Because of the manner in which the collection was acquired, the provenance of many objects is uncertain and a significant number of the records are erroneous. Consultation with Native community members has provided a means of correcting the catalog information as well as providing a much deeper knowledge base on the cultural significance of the collections. The virtual museum projects provide another form of consultation that serves that purpose.

The essay by Galvin Martinez on his great-aunt's pottery plate cited earlier vividly illustrates the kind of contribution that virtual museum participants can make to the museum's information on its collections. The following examples underscore this point.

The catalog information for one of the Yup'ik masks in one project was unusually complete compared to most catalog entries. Few of Heye's entries are more detailed than this one:

> Mask representing Negakfok, the cold weather spirit. This spirit likes cold and stormy weather. It is used in the dances in March and has a sad expression on account of the approach of spring.
>
> Eskimo
>
> Near Bethel, Kuskokwim River Alaska.
>
> A. H. Twitchell Collection

Compare the catalog entry to the virtual museum text label created by the Yup'ik students:

> This mask is from the village of Napaskiak, seven miles down the Kuskokwim River from Bethel, Alaska. The owner of the mask is unknown, but it is certain that he/she was a shaman.
>
> During the winter, Yup'ik people gathered together in the local men's house to perform a ceremony called *Agayuyaraq* or *Itruka'ar*. In this ceremony, people asked *Ellam Yua*, The Creator, for abundance of animals and good weather for hunting.
>
> The local shaman designed the mask, which depicted his or her *tuunraq*, helper spirit. The shaman also composed the song and choreographed the dance that would accompany the mask.
>
> The song usually contained a chorus and two verses. This mask represents the wind that brings cold weather in winter, ideal for hunting and fishing. This mask also has a pointed forehead, which

> probably represents the shaman's helper spirit; *ircenrraq*, an underground supernatural being. These two were included in the lyrics of the song.
>
> Before the arrival of white men, Yup'ik people made paint using natural pigments. There were four primary colors: black, red ochre, white, and blue. The maker of this mask used black paint, *avisgaq* in Yup'ik. The decorations at the top of the mask dangled with the movement of the dancer, making the sound of the blowing wind. The eye balls in the center of the sockets rolled with the movement of the dancer. The four fingered hand means that the mask's owner was a shaman.
>
> Masks like this one were used only once and were either burned as fire wood in a fire bath or left in the tundra, far from the reaches of man.
>
> *–Elsie Alexie and Nicholai Joekay; Napaskiak, AK*

In comparison, the catalog entry comes off as a dry academic description from an anonymous source. The original author probably had some Native informant, but the information that was solicited demonstrates little understanding of Yup'ik culture. The students clearly identify themselves as residents of the village the mask originally came from, and they add to the specificity of that location. They project a special authority to their report, giving rich contextual detail, replete with Native language terms. The students identify their culture as Yup'ik rather than Eskimo, a common kind of correction that Native people have been making to the names and categories that have been imposed upon them by Western academics.

A second example also comes from the Yup'ik project. The item is a pair of feathered dance hoops. The catalog entry says:

> A pair of wooden hoops decorated with feathers held in hands during dance.
>
> Eskimo
>
> Good News Bay, Alaska
>
> Purchase

For this item, the students produced no text label. Instead, they were videotaped giving spontaneous performances of their village dances using the dance fans. The performances included the accompanying songs in the Yup'ik language in beautiful harmony. The students were wearing blue jeans and tee shirts, emphasizing that they were representing their living culture in a contemporary context. The students and adult participants were so delighted with the opportunity to dance with these fans–probably for the first time in a hundred years–that they performed one song after another. One of the adult participants pointed out, "Our children grow up with their music." Though this kind of use of fairly fragile objects would be prohibited by standard museum practice, it was deemed worth the risk of damage to reunite the items in this manner with the people who possess such a strong cultural connection to them. Museum conservators enjoyed the demonstration along with everyone else.

After about a half dozen dances, the fans were put away, so that the virtual museum project could move on to other objects selected for that day's work. But the singing continued, like an ongoing conversation about the wonderful experiences of the day. After awhile, non-Yup'ik words could be heard mixed in the songs, such as "NMAI" and "Union Station." Perhaps this virtual museum project will become part of the Yup'ik village repertory.

A FORUM FOR CULTURAL STORYTELLING

W. Richard West, the Southern Cheyenne director of the National Museum of the American Indian, has advocated a departure from the traditional concept of the museum as a temple of authoritative knowledge and a move toward the concept of the museum as a forum for multiple worldviews, where truth is relative and has diverse manifestations. He characterizes the prior museum practice through the manner it presented Native people. This older model relies "almost exclusively on voices of third parties or those who in reality sit outside the Native cultural experience and who often approach it based upon academic disciplines or systems of knowledge that are Western rather than Native in origin." The approach at the NMAI, however, has been "very much a forum that encourages and assumes a multiplicity of authoritative viewpoints in interpreting Native cultures and peoples. It is a museum where what the visitor reads, sees, or hears about an object, cultural or historical proposition, or cultural phenomenon may vary depending upon who is speaking."[37]

The NMAI has been implementing this transformation in museum practice through many types of Native community consultations. In this paper we have focused on one particular type of consultation, projects in which Native students use 21st century technologies and conduct research employing secondary and primary sources to create virtual exhibitions featuring museum objects relating to their cultures. The practice has had important, beneficial effects for the students, their communities, and both public and tribal museums. It gives the participants the opportunity to access important cultural objects and tell the stories of those objects in new ways using new information technologies. In the process, the students adopt the roles of culture bearers engaging in intercultural communication in a way that challenges stereotypes, corrects misinformation, and represents living, vital cultures with deep connections to the past. This model is a forum for telling cultural stories, both old and new, stories that may then be shared–responsively and responsibly–on the World Wide Web.

NOTES

1. Mark Christal, "Virtual Museum Projects for Culturally Responsive Teaching in American Indian Education" (Ph.D. diss., University of Texas at Austin, 2003). Also Mark Christal, Loriene Roy, Paul Resta, Martha Kreipe de Montano, and Antony Cherian. "Virtual Museum Collaborations for Cultural Revitalization: The Four Directions Model." Paper presented at the Museums & the Web Conference, Seattle, March 16-19, 2001. [Accessed 19 April 2004] Available at URL: http://www.archimuse.com/mw2001/papers/christal/christal.html.

2. Garrick Mallery, *Picture Writing of the American Indians* (New York: Dover, 1972).

3. W. Richard West, "Schools Can Now Access National Museum of American Indian Online," *Indian Country Today*, March 29, 2000.

4. C. Grill, "Delta Students Create Virtual tour of Yup'ik Cultural Artifacts," *The Delta Discovery* (Bethel, Alaska), 1 October 2000, 17.

5. Grill, "Delta Students," 17.

6. "Santa Clara Pueblo Students Create Virtual Museum Tour." *NMAI Runner* (1999) 99 (2): 6.

7. Smithsonian Institution, National Museum of Natural History, Repatriation Office. "National Museum of the American Indian Act and Amendment." [Accessed 4 April 2004] Available at URL: http://www.nmnh.si.edu/anthro/repatriation/nmaiact.htm.

8. United States Department of the Interior, Bureau of Reclamation. "Native American Graves Protection and Repatriation Act." [Accessed April 4, 2004] Available at URL: http://www.usbr.gov/nagpra/naglaw.htm.

9. S.C. Kinser, "Museums and Tribes: A Tricky Truce," *New York Times,* December 24, 2000, Section 2, 1, 39.

10. W. Richard West, Speech and interview given to the Washington Press Club, 15 January 2004.

11. Nancy B. Rosoff, "Integrating Native Views into Museum Procedures," *Museum Anthropology* 22 (1998): 33-42.

12. Alaska Native Knowledge Network. "Guidelines for Respecting Cultural Knowledge." Adopted by the Assembly of Alaska Native Educators 1 February 2000. [Accessed April 14, 2004] Available at URL: http://www.ankn.uaf.edu/standards/.

13. Jane Hunter, Bevan Koopman, and Jane Sledge. "Software Tools for Indigenous Knowledge Management." Paper presented at the Museums & the Web Conference, March, 2003. [Accessed April 14, 2004] Available at URL: http://www.archimuse.com/mw2003/papers/hunter/hunter.html.

14. Christal, "Virtual Museum Projects."

15. Christal, "Virtual Museum Projects," 154.

16. Aliases of persons and places were used in the Christal dissertation to provide confidentiality to his subjects. The tribal associations did indicate the regional identification of the cases, e.g., Pacific Northwest tribe.

17. Christal, "Virtual Museum Projects," 147.

18. Christal, "Virtual Museum Projects," 352.

19. Christal, "Virtual Museum Projects," 153.

20. Christal, "Virtual Museum Projects," 154.

21. Christal, "Virtual Museum Projects," 305.

22. Christal, "Virtual Museum Projects," 307.

23. Christal, "Virtual Museum Projects," 442.

24. Christal, "Virtual Museum Projects," 240.

25. Christal, "Virtual Museum Projects," 442.

26. Christal, "Virtual Museum Projects," 375.

27. Christal, "Virtual Museum Projects," 447.

28. Christal, "Virtual Museum Projects," 182.

29. Christal, "Virtual Museum Projects," 181.

30. Vivian Delgado, "Technology and Native America: A Double-Edged Sword," in *Toward Digital Equity: Bridging the Divide in Education*, edited by G. Solomon, N. Allen, and P. Resta (Boston: Allyn and Bacon, 2003), 89-95.

31. Christal, "Virtual Museum Projects," 494-5.

32. Christal, "Virtual Museum Projects," 495.

33. Christal, "Virtual Museum Projects," 494.

34. Christal, "Virtual Museum Projects," 497.

35. Christal, "Virtual Museum Projects," 497.

36. W. Richard West, Speech and interview given to the Washington Press Club, 15 January 2004. For further information on Heye, see Lawrence M. Small, "A Passionate Collector." *Smithsonian Magazine,* November 2000. [Accessed 14 April 2004] Available

at URL: http://www.smithsonianmag.si.edu/smithsonian/issues00/nov00/small_nov00.html and National Museum of the American Indian. "Biography: George Gustav Heye." [Accessed 11 April 2004] Available at URL: http://www.nmai.si.edu/press/releases/gghc_biography.html.

37. W. Richard West, "The National Museum of the American Indian: A Prototype for the 21st Century?" *Archiv für Volkerkunde* 50 (1999): 115-123, 119.

Chinese Collections in Museums on the Web: Current Status, Problems, and Future

Hsin-liang Chen

SUMMARY. This paper focuses on types of images indexed by museum practitioners, the indexing procedures and elements, and types of tools and systems used. The six participating museums are in the states of California, Illinois, Massachusetts, New York, Ohio, and Washington, DC. Their conventional cataloging and indexing practices are not suitable or transferable for the new image management system. The lack of indexing standards and tools is the common challenge faced by the six museums involved in the study. Most image management systems are not metadata/XML ready and the expansion of the systems onto the Web may be limited, which contributes to the internal conflicts that exist within the museums. *[Article copies available for a fee from The Haworth Document Delivery Service: 1-800-HAWORTH. E-mail address: <docdelivery@haworthpress.com> Website: <http://www.HaworthPress.com> © 2004 by The Haworth Press, Inc. All rights reserved.]*

Hsin-liang Chen is affiliated with the School of Information, University of Texas at Austin, 1 University Station, D7000, Austin, TX 78712 (E-mail: chen@ischool.utexas.edu).

The author gratefully acknowledges the support and assistance of the museum practitioners at the six museums.

This project is supported by the University of Texas at Austin's 2002 Faculty Summer Research Assignment and the School of Information's Temple Teaching Fellowship.

[Haworth co-indexing entry note]: "Chinese Collections in Museums on the Web: Current Status, Problems, and Future." Chen, Hsin-liang. Co-published simultaneously in *Journal of Internet Cataloging* (The Haworth Information Press, an imprint of The Haworth Press, Inc.) Vol. 7, No. 1, 2004, pp. 89-102; and: *Collaborative Access to Virtual Museum Collection Information: Seeing Through the Walls* (ed: Bernadette G. Callery) The Haworth Information Press, an imprint of The Haworth Press, Inc., 2004, pp. 89-102. Single or multiple copies of this article are available for a fee from The Haworth Document Delivery Service [1-800-HAWORTH, 9:00 a.m. - 5:00 p.m. (EST). E-mail address: docdelivery@haworthpress.com].

http://www.haworthpress.com/web/JIC

Digital Object Identifier: 10.1300/J141v07n01_06

KEYWORDS. Image management, Chinese museum collections, museum Web sites

INTRODUCTION

With the development of computing technology, many Chinese museums and museums with significant Chinese collections have digitized and provided images of their collections on Web sites. These online resources offer users around the world access to valuable treasures to learn more about Chinese culture. However, there are obstacles that must be overcome to achieve the goals of promoting the Chinese heritage and educating new generations.

The purpose of this project is to study how museum practitioners use current image indexing practices and services to retrieve the images of the Chinese collections. Several issues, including image needs, information-seeking strategies, information queries, search functions, display formats, and human-computer interaction are examined in this study.

This paper focuses specifically on the current practices of image management: types of images indexed by museum practitioners, the indexing procedures and elements, and types of tools and systems used.

BACKGROUND OF THE PROBLEM

Current Image Management at Museums

Dervin and Nilan[1] point out that user search behavior analysis and specific domain applications need to be addressed to further enhance studies in information retrieval systems. Su[2] suggests that the development of effective information retrieval systems should rely on systematic feedback from evaluation by real users with real information needs. Graham[3] surveyed 60 art libraries in the U.K. The survey included the important issues of image collections, cataloging and indexing practices, content-based image retrieval (CBIR) systems, and the use of images. Graham's study reports on the current management of image collections and techniques for image and video retrieval in the U.K. Eakins and Graham[4] study the current state of the art in CBIR systems within the U.K. and submit several suggestions to U.K. governmental agencies, users, and managers of image collections and CBIR software developers. The Visual Image User Study (VIUS) project at The Pennsylvania State University is conducting an extensive and systematic assessment of its needs for digital image delivery.[5] The VIUS project is working to develop digital picture libraries to supply new uses of digital images for teaching and research.

Importance of Information Needs and Information-Seeking Behavior to System Design

Some previous studies have shown that museum visitors liked to browse exhibits; their behavior was more like sight-seeing or window-shopping.[6-7] Davies[8] reported that museums and galleries in the United Kingdom have been required by the government to provide audiences with access to their collections culturally and intellectually. He also states that Web pages as marketing tools present information about exhibitions, taxonomies, and organizing display principles of objects which can provide motivation for visiting (p. 286). Cameron[9] stated that thematic interfaces to museum collections have been more popular and have shown an important paradigm shift (p. 309).

Stephenson[10] examined several cultural heritage image databases and identified key issues for future improvements. She pointed out that, in addition to technological challenges, the areas of audience, user behavior, and use should be addressed as well. Dyson and Moran[11] studied seven Web sites (museums, libraries, galleries, educational projects, photographic collections) and found five of the seven Web sites provided a searchable database of their collections with limited search functions (p. 396). Different tools may be required by special users and environments. New functions may be created to facilitate users' search strategies. To achieve such goals, studies on interface design, human-computer interaction, and users' information-seeking behavior should be conducted. Davies[8] suggested that "information about, and physical and intellectual orientation and navigation to, collections can be provided through a media programme using symbols," and a prototype was created at Birmingham Museum and Art Gallery (p. 283). However, there is no evaluation of the prototype available.

Other Challenges

In addition to the above key issues, the museums also face several challenges: (1) lack of communication among museums; (2) lack of indexing standards and tools; and (3) lack of translation standards of Chinese into western languages. Chinese museums and museums with significant Chinese collections should form a consortium to establish communication and to develop collaboration. Many western museums have begun those efforts. The Art Museum Image Consortium (AMICO) is one of those consortiums, but its image collections have only about 6,000 works from Asian cultures.[12] The lack of indexing standards and tools presents the same challenge for all museums. Most museums either develop their own indexing standards and tools or do not have adequate professional personnel to manage their image collections.[3,13] Regarding translation standards, although American libraries started using Pinyin as the standard romanization scheme for Chinese characters on October 1, 2000,[14] many museums may not be aware of this change and may still use the Wade-Giles system. These challenges are important to the development of image collections.

RESEARCH QUESTIONS

This paper focuses on an initial study on image management at Chinese museums and museums with significant Chinese collections in the digital age. The following research questions were investigated:

- What kind of images do the museums index?
- How do the museums index their image collections?
- What kind of indexing tools do the museums use?

METHODOLOGY

Participants

Six museums were selected for this study based on the size and diversity of their Chinese collections or their image management. The six participating museums were in the states of California, Illinois, Massachusetts, New York, Ohio, and Washington, DC.

Procedure

Pre-Visit Questionnaires

Graham's survey was adopted in this project.[3] A set of self-administered questionnaires was used to collect image managers/slide librarians' views on cataloging/indexing practices, the functions of new image management systems, and the use of images. The questionnaires were distributed to correspondents before an on-site visit and were collected between January and April 2002. The six museums have different internal administrative structures and their various offices are handling the transformation of image digitization. Due to the multiple layers of complexity, the terms "slide librarians" and "image managers" are interchangeable throughout the entire paper unless otherwise noted.

Follow-Up Phone Interviews

After all respondents answered the questionnaires, the investigator examined the questionnaires and conducted phone interviews with the respondents for unclear answers and in-depth information. The investigator identified several key people for observations and interviews when visiting museums.

On-Site Visits

Based on the knowledge gained from the questionnaires and phone interviews, on-site visits were conducted between June and August 2002. The investigator observed librarians' and museum practitioners' image-seeking behavior and also interviewed those people for further understanding of their search behavior. The investigator interviewed museum administrators to obtain their expectations for digital image management in the mission of the museum.

RESULTS

The six participating museums are currently establishing a database serving areas such as the registrar office, photo studio, curatorial departments, education/outreach programs, and library. The purpose and the scope of the database vary in the six museums. In general, each database consists of information about the museum collections including imagery. The investigator reports the results collected from the surveys, on-site visits, and interviews with regard to the three research questions.

Research Question 1: What kind of images do the museums index?

Fourteen types of images were surveyed: photographic prints, photographic negatives, transparencies/slides (35-mm), transparencies (5″ × 4″), posters, fabrics, film, video, paintings, prints, drawings, art productions, published scrolls, and digital images. The majority of the analog image collections at the six museums includes photographic prints, photographic negatives, 35-mm slides, and 5″ × 4″ transparencies. The six museums have digitized their image collections using different strategies and timeframes. One museum has digitized over 5,000 photographic prints, another has digitized over 100,000 photographic negatives, another has digitized over 5,000 35-mm slides, and one has digitized around 10,000 transparencies.

During the on-site visits, the investigator found that the estimations of the image collections from the surveys would not be comparable since the leading offices of the digitization process at the six museums were very different. Three museums had their information technology (IT) departments manage the digitization process, while the others used the registrar office, the photo studio, and the project office (a university museum). Each of the six museums has a slide library; the six slide libraries house different image collections reflecting the missions and policies of the museums. However, the quality of the slide collections is not good for digitization. Most slide libraries host slides primarily and receive analog images in different formats from the photo studio, since the collaboration between the slide library and the photo studio is not consistent at the same museum over time. Due to the inconsistencies, the slide libraries do not have complete sets of images of the museums' collections.

Missing analog images are common among the six museums. Currently, the six museums have started checking their inventories. The investigator could receive only rough estimations of the analog and digital images from the slide librarians and image managers.

Copyright issues are another consideration, since the slide libraries also receive 35-mm slides from other museums and art galleries. Some slide libraries even take slides from printed materials upon requests from users. Some slide libraries duplicate the slides for their library patrons who may be museum practitioners or outside members. These different practices complicate the study on the size of the analog image collections.

Research Question 2: How do the museums index their image collections?

Different departments at the six museums try to figure out how to index their image collections with the use of the registrar database. Table 1

TABLE 1. Types of Descriptors

Descriptors / Type of Material	*ID number*	*Author, photographer, source, etc.*	*Title, name*	*Physical description (e.g., size, length)*	*Content Descriptor*	*Place names*	*Date*	*Other*	*Not described formally*
Photographic prints	3	3	3	3	1	2	3		
Photographic negatives	4	4	2	3	1	1	3	1	
Transparencies/slides (35-mm)	4	2	2	3	1	2	3	1	
Transparencies (5″ × 4″)	3	3	1	2	1	1	2		
Posters									
Fabrics									
Film									1
Video	1		1	1	1	1	1	1	1
Paintings	2	2	2	2	1	2	2		
Prints	2	1	1	1	1	1	1		
Drawings	2	2	1	2	1	1	2		
Art reproductions									
Other	1	1	1	1	1	1	1	1	1

shows that different types of analog image collections are indexed with different recording elements in the current practices at the slide libraries. The popular descriptors are ID numbers, authors, titles, physical descriptions (e.g., size, length), place names, and date.

The current practice inherits the generational difficulties and limits. Most slide librarians could not update descriptors due to the limited personnel and resistance from library users and museum practitioners. According to the interviews, the investigator found that the slide libraries did not have adequate support on the descriptive and subject analyses for their analog image collections. On the other hand, the library users are rarely aware of the importance of the descriptive and subject analyses and they most often rely on the slide librarians' assistance to find images. Such poor practices and misconceptions hinder a transition from the current indexing practice to the new registrar database.

Translation standards also affect the selection of the indexing terms. The museum libraries have converted their bibliographic records to the Pinyin system with assistance from OCLC. On the other side, since the slide collections and their records are still in the analog form, most records are still in the Wade-Giles system.

Research Question 3: What kind of indexing tools do the museums use?

In-house indexing tools are commonly used in the six museums. AACR II and MARC are used by three museums for selected materials only. The popular content descriptors are title, period, genre, subject headings, and classification scheme. Regarding classification schemes, most museums use their in-house rules. Only one museum uses the *Library of Congress Classification*, the *LC Subject Headings*, and *Art & Architecture Thesaurus* for most of its image collections. Another museum uses the *LC Subject Headings*, and *Art & Architecture Thesaurus* for its 35-mm slides only.

The in-house tools and standards are products of long-standing practices, analog images, limited access, and inadequate personnel. Some slide librarians/image managers stated that most in-house rules and standards were developed before they took their posts.

Basically, the function of the slide libraries is to facilitate museum practitioners in performing their job duties, such as arranging exhibitions, publishing scholarly papers, planning educational programs, etc. Most slide libraries are also open to the public. In the physical environment, slide libraries provide in-person services to the user. Such services satisfy users' needs and minimize the insufficiency of indexing tools and standards. The current tools and standards are not transferable to the digital image collections.

In addition to the three research questions, the investigator discovered several internal conflicts in the museums from the on-site visits.

Conflict 1: Record Management

The first difficulty is to match physical objects with their images and registrar records. The mission of the digitization at the six museums is to establish a comprehensive registrar database consisting of a record of each object and at least one image of the object. However, the six museums do not have complete records of their collections. For example, some museums have records without any image of the objects while some museums have only brief records of the objects after purchasing them. In order to create such a database, the registrar office, the photo studio, and the IT department must work together. The registrar office has to clean up and update the records; the photo studio needs to produce images for each object, and the IT department is responsible for designing/purchasing an image management system. All six museums have initiated the design of their registrar database. Some of them have cleaned up the registrar records while some are digitizing substantial amounts of digital images. The strategies adopted by the museums to create their database are very different and they are somehow problematic.

Some photo studios are scanning the existing analog images to improve the quality of the digital copies with image-editing software. Some photo studios are scanning available analog images and taking new photos of the collections in the digital format only. The problems of the image collections are:

- Establishing records: The six museums are re-establishing the records of their collections. Inadequate records and missing or aged images are common situations at the six museums.
- Image acquisition: This is a technical and philosophical problem. Some curators insist that current photographic technology and digital devices do not display quality images on the computer monitor. Photographers' reactions to curators' comments are "That is just their psychological bias." Different views on the digital image acquisitions may have a snowball effect in the future. The number of images is also a problem. None of the six museums have policies on the determination of the number of images to be taken for each object. At some museums the photo artists can decide how many images they should take for one object, while curators decide the number of images and the details of the images at another museum. At one museum, only one official image can represent one object and the image must be approved by the curator.
- Image marketing: The museums are spending a considerable part of their budgets on image acquisition. However, they seldom consider selling analog or digital images to generate revenues. The expenses of recovery and maintenance of state-of-the-art photo facilities challenge each museum.

Conflict 2: Administrative Responsibilities

The six museums employ different record-keeping elements of their objects. At some museums the registrar office is in charge of the digitization project, so the recording elements may lean toward that particular department's interest. This is a common trend in the museums. Some museum practitioners expressed their confusion and frustration due to the complexity of jargons in use among the different departments. The completion of one record requires the involvement of several departments. For example, the property information is from the registrar office; the image comes from the photo studio (at some museums the curators have to approve the quality of the image); and other descriptions come from the curatorial departments or the slide library.

Curatorial departments hold higher administrative positions at some museums and they are more influential in decision-making. Some of them do not consider participating in the creation of the registrar database a top priority. Such an attitude hinders the progress of the registrar database, since the curatorial departments are responsible for inputting indexing terms and detailed descriptive information, and approving the accuracy of each record including images of the objects. Such resistance from the curatorial departments is understandable, because many curators do not have the knowledge and skills to deal with computer-based databases and indexing, but they are ultimately responsible for the quality of the records. Due to their professional beliefs, they process each record precisely. However, the process has caused some practical problems. For example, one museum purchased the same image management system for its curatorial departments; the curatorial departments customize the system independently and the systems are not linked to each other.

Conflict 3: Purpose of the Database

The six museums have not decided to what extent the registrar database can serve the museum practitioners (internal users) and museum visitors (external users) better, and how to derive information from the registrar database for different users. The registrar database is huge and should host comprehensive information about the museums' collections. However, each department has different concerns in regard to how an object should be described, specifically curatorial departments that are dealing with different objects and cultures. It is a great challenge to coordinate all the different departments. Internally, the registrar database can assist museum practitioners to find information about museum collections to perform job duties. Externally, museum visitors and scholars are interested in finding information for pleasure and research. Satisfying these different needs is difficult. So far, major efforts at the six museums are devoted to making basic record-keeping elements (i.e., metadata) and one image of the object available.

Treinen[6] and Semper[7] pointed out that museum visitors' behaviors was more like sightseeing and window-shopping. The visitors did not come to the museums in a planned manner. The studies were concerned about the museum visitors in the physical environment. How about visitors to museums' Web sites? Do online visitors also randomly browse the images of the collections? Do they really care about any sophisticated browsing/search functions for viewing the images? Most museums do not know who their online visitors are. This question must be answered before determining the indexing tools and standards. The use of the indexing tools and standards should aim to facilitate different users' behavior.

Even though thematic interfaces to museum collections have been more popular and shown an important paradigm shift,[9] one approach will not satisfy all of the different user groups (p. 309). Search interfaces, indexing tools, and standards must be integrated into the design of the registrar database. Such integration varies from museum to museum.

Conflict 4: Standards and Database Structure

Some museums have tried to accommodate some emerging standards such as Encoded Archival Description (EAD), the Computer Interchange of Museum Information Consortium (CIMI) standards, and the Visual Resources Association's VRA Core, etc. However, the adoption of these emerging standards is relative to each museum's existing system. Many systems already have built-in standards for data descriptions. Some systems provide limited flexibility for customization. The common problems are:

- Multiple layers of descriptions: Some systems have four to five layers for each record. Maintaining these records is time consuming and exhausts museum practitioners' energy.
- Professional jargons: Many practitioners are not familiar with the names of record-keeping elements and their meanings in the systems even though they are responsible for data entry.
- Cross-cultural consistency: The emerging standards may be more suitable for Western cultural objects but not for non-Western cultural objects in several areas such as translation and retrieval.
- Multiple languages: Some systems cannot process foreign languages.

- Authority control for indexing: Disagreement on the use of the indexing terms often occurs between the slide library and the curatorial departments. In the physical environment, most curatorial departments have their own slides and have arranged the slides in their preferred fashion. In the digital environment, such practices have to be changed.

DISCUSSION

This research identified the following problems with the development of image management systems in six museums which have digitized their Chinese collections. All these museums have used or purchased a computer-based image management system to organize the digital images and their related information. The most popular formats of the original images are photographic prints, photographic negatives, and 35mm slides.

Most museums did not have comprehensive records in the past, so they have spent substantial amounts of their budgets and personnel to establish basic records or re-enter data for the digital images. Their conventional cataloging and indexing practices are not suitable or transferable for the new image management system. Image management systems used by the six museums are not able to accommodate the features of the Chinese collections and their records. Currently, the six museums should address the following problems:

- Cross-board indexing standards: Existing standards such as *Dublin Core*, *EAD*, *VRA Core*, etc., do not satisfy every department's job requirements. Each museum must have policies on the indexing standards that can serve the different departments and will not over-burden the staff.
- Museum-wide training: Museum practitioners should receive training on image digitization, computer database management, and indexing, since they are responsible for inputting information into the registrar database. The successful implementation of the registrar database relies on the quality of museum practitioners' work. Education is the key to the success.
- New reward system and job requirements: A reward system and new job requirements should be established to honor the staff's participation and performance. One museum has already imple-

mented such a procedure. The new reward system will motivate the staff to produce quality information and the job requirements will require that the staff carry out the tasks.

- Database and workflow: The design of the registrar database should improve the workflow among departments. The database has to monitor the workflow of data entry from each department in a timely manner. On the other side, the workflow regulates the museum practitioners' job duties.
- Image quality and acquisition: The six museums should have policies on image quality and acquisition. Digital images can be used for publishing and display in different environments and they have different file formats with various qualities. Producing digital images in different file formats and managing them in the registrar database are a great challenge. Some current management systems can handle certain file formats only and need to display the images with more functions such as zoom-in/out, rotation, and scaling.

At the current stage, the six museums need to solve these above mentioned internal problems. Only then may they extend their services to external users. Some museums' administrators have recognized some of the problems and require museum practitioners from different departments to report their progress regularly.

Most image management systems are not metadata/XML ready, which means that the expansion of the systems onto the Web may be limited. The development of indexing schema is critical to the management of digital images and to the museum practitioners and online users. Image management systems should be enhanced with the standards of metadata/XML, etc., for the Web-based environment.

One important issue is that the museums need a systematic approach to serve their external users. Such an approach has not been planned at most museums since the museums are focusing on their internal users. Museum visitors, scholars, and educators look for information differently. The museums are facing the following challenges: understanding the external users' differences and deriving information from the registrar databases to serve these different users. The museums need to evaluate their available personnel and resources for such services. Should educational/community outreach programs come first or detailed scholarly materials? They need to know how to market their digital resources and set up the priorities.

REFERENCES

1. Brenda Dervin, and Michael Nilan, "Information Needs and Uses," *Annual Review of Information Science and Technology* 21 (1986), 3-33.

2. Louise T. Su, "Developing a Comprehensive and Systematic Model of User Evaluation of Web-Based Search Engines," in *Proceedings of the 18th National Online Meeting*. (New York, 1997), 335-44.

3. Margaret E. Graham, "The Description and Indexing of Images: Report of a Survey of ARLIS Members, 1998/99," [Accessed 15 November 2003] Available at URL: http://www.unn.ac.uk/iidr/ARLIS/.

4. John P. Eakins and Margaret E. Graham, "Content-based Image Retrieval: A Report to the JISC Technology Application Programme," [Accessed 15 November 2003] Available at URL: http://www.unn.ac.uk/iidr/research/cbir/report.html/.

5. Henry Pisciotta et al., "Penn State Visual Image User Study," *D-Lib Magazine* 7: 7/8 (2001) [Accessed 15 November 2003] Available at URL: http://www.dlib.org/dlib/july01/pisciotta/07pisciotta.html/.

6. Heiner Treinen, "What Does the Visitor Want from a Museum? Mass Media Aspects of Museology," In *Museum Visitor Studies in the 90s* (London, Science Museum, 1993), 86-93.

7. Robert J. Semper, "Designing Hybrid Environments: Integrating Media into Exhibition Space," In *The Virtual and Real: Media in the Museum* (Washington D.C.: American Association of Museums, 1998), 119-27.

8. Roderick Davies, "Overcoming Barriers to Visiting: Raising Awareness of, and Providing Orientation and Navigation to, a Museum and Its Collections Through New Technologies," *Museum Management and Curatorship* 19 (2001): 3, 283-95.

9. Fiona Cameron, "World of Museums: Wired Collections–The Next Generation," *Museum Management and Curatorship* 19 (2001): 3, 309-15.

10. Christie Stephenson, "Recent Developments in Cultural Heritage Image Databases: Directions for User-Centered Design," *Library Trends*, *48* (1999): 2, 410-37.

11. Mary C. Dyson, and Kevin Moran, "Informing the Design of Web Interfaces to Museum Collections," *Museum Management and Curatorship* 18 (2000): 4, 391-406.

12. Art Museum Image Consortium (AMICO). [Accessed 15 November 2003] Available at URL: http://www.amico.org/.

13. Council on Library and Information Resources (CLIR), *Scholarship, Instruction, and Libraries at the Turn of the Century: Results From Five Task Forces of Learned Societies and the Council on Library and Information Resources* (Washington, D.C.: Council on Library and Information Resources, 1999).

14. Research Libraries Group (RLG) "RLG Forum: Wade-Giles/Pinyin Conversion Planning." 2001. [Accessed 15 November 2003] Available at URL: http://www.rlg.org/eas/pinyinforum.html/.

Patterns of Identification of Potentially Sensitive Data in Natural History Museum Online Catalogs

Bernadette G. Callery

SUMMARY. Staff in natural history museums typically protect sensitive information held in collection records by mediating access to collection data. Such information includes collection location data, particularly of endangered species, and is generally supplied to users requesting it on an individually determined need-to-know basis. This study examined the online collection catalogs of 23 North American natural history museums, focusing on statements concerning the specific identification and availability of sensitive data when releasing collection information in this unmediated environment. Only 8 of the 55 separate databases surveyed during November 2003-February 2004 were found to include a specific identifying statement of sensitive data and its inclusion or exclusion in the museum's online collection databases. *[Article copies available for a fee from The Haworth Document Delivery Service: 1-800-HAWORTH. E-mail address: <docdelivery@haworthpress.com> Website: <http://www.HaworthPress.com> © 2004 by The Haworth Press, Inc. All rights reserved.]*

Bernadette G. Callery, PhD, is Museum Librarian and Head, Library and Archives, Carnegie Museum of Natural History, 4400 Forbes Avenue, Pittsburgh, PA 15213 (E-mail: calleryb@CarnegieMNH.org).

[Haworth co-indexing entry note]: "Patterns of Identification of Potentially Sensitive Data in Natural History Museum Online Catalogs." Callery, Bernadette G. Co-published simultaneously in *Journal of Internet Cataloging* (The Haworth Information Press, an imprint of The Haworth Press, Inc.) Vol. 7, No. 1, 2004, pp. 103-115; and: *Collaborative Access to Virtual Museum Collection Information: Seeing Through the Walls* (ed: Bernadette G. Callery) The Haworth Information Press, an imprint of The Haworth Press, Inc., 2004, pp. 103-115. Single or multiple copies of this article are available for a fee from The Haworth Document Delivery Service [1-800-HAWORTH, 9:00 a.m. - 5:00 p.m. (EST). E-mail address: docdelivery@haworthpress.com].

http://www.haworthpress.com/web/JIC

Digital Object Identifier: 10.1300/J141v07n01_07

KEYWORDS. Digital museums, virtual museums, sensitive data, collection management systems

INTRODUCTION

The public's trust in museums is based on the museums' ability to responsibly manage both their collections and the information about them. Public access to that information is an obligation stipulated in the American Association of Museums Code of Ethics which states that "the stewardship of collections entails the highest public trust and carries with it the presumption of rightful ownership, permanence, care, documentation, accessibility and responsible disposal."[1] Prior to the widespread use of online public access catalogs to museum collections, museums provided mediated access to information about objects in their collections through exhibition labels and catalogs or by direct inquiry to the museum's curators. These techniques allowed museum staff to control access to sensitive information and to assemble and interpret existing collection records to meet the needs of the specific user. Arguments in support of this mediation include the need to select and interpret documentation about specimens and artifacts in the collection, which historically can vary widely in veracity and levels of detail. Raw data found in collection records such as accession ledgers, specimen tags, field notes, and correspondence, may require elaboration, and archaic or vague geographical references may require further research in order to establish present-day equivalents.

Museums may define sensitive information as the name of the donor, the amount of the purchase price, or the location where the object is stored within the museum. Natural history museums with examples of endangered species in their collection, or archaeological or paleontological material, generally consider the location where the specimen was originally found or collected to be sensitive information. The primary argument for limiting access to detailed locality data is the protection of economically exploitable objects from over-collecting. Such collecting can result in both habitat destruction and loss of species, as well as loss of context or strata data when objects are removed from an inadequately documented archaeological or paleontological site.

While restricted access to collections and their associated data have been historically managed through one-on-one correspondence or in situ research visits, the purpose of this study was to determine what restrictions were in effect in the unmediated environment of online collection catalogs.

METHODOLOGY

In order to investigate the treatment of sensitive information in publicly accessible collection databases, twenty-three United States and Canadian natural history museums were selected and their Web sites examined during November 2003 to February 2004 for evidence of statements of the identification and inclusion or exclusion of sensitive data in their online collection catalogs.[2] Museum selection was based on a recent benchmarking study undertaken by the Carnegie Museum of Natural History to investigate the availability of online collection information in similar institutions.

For purposes of this study, an online collection database was defined as any user-searchable tool in which the user could select one or more index terms in order to locate information about the presence of a specific object in the museum's collection. Sites with collection databases were further examined for statements of completeness of the searchable data or statements of exclusion of specific types of sensitive information. Collection information presented in a narrative or extended discussion of the collection's contents, historical development, major collectors, or donors was not considered in this study, nor were lists or counts of specimens by species or geographical distribution. This study concentrated on one aspect of the general information dissemination problem: the nature of statements identifying how sensitive information is handled in online access to natural history museum collections.

FINDINGS

Even with the availability of online collection catalogs, the tradition of non-integrated museum collections continues to predominate. Access to museum collection information is largely limited to searching in individual discipline-based collection catalogs. Few museum-based catalogs allowed searching across all available collections in a given museum. A few sites included portals which allowed users to search across collections in other museums, but these were limited to collections in specific disciplines.

The survey found 125 separate museum collections represented on the Web sites of the 23 institutions, with 55 individual collections available for unmediated searching of the collection databases. The disparity between the number of institutions and the number of museum collection sites is due to the fact that most natural history museums still maintain separate collection catalogs for objects in individual scientific disciplines. Eight of these collection databases included statements of exclusion of sensitive information either of entire records of selected groups of objects or of specific fields of information from the specimen record. In all noted cases, this sensitive information involved data on the location of endangered and threatened plant and animal material. Data on these species in the museum's collection was either suppressed entirely from the database or only partially reported. Users wishing additional or more specific information were invited to apply directly to the appropriate curator, where their request would be considered on an individual basis, following the pattern of collection access established in the pre-electronic era.

The New York Botanical Garden's site included the most extensive discussion of the rationale for excluding location information on endangered species. Their "Brief Note Concerning Specimen Records of Endangered and Threatened Plant Species" explicitly states their suppression of locality data for such plants. Furthermore, the site applies those same limitations to all species from groups considered typically overcollected.

> In an effort to meet our obligations to protect populations of endangered species from over-collection, some of our specimen data (such as specific locality information) have been removed from those online records for endangered species. These data are made available to researchers on request. We are aware that making our herbarium specimen data available involves striking a delicate balance between access to data that are important to research and the potentially reckless posting of sensitive information. To that end, we are in the process of removing portions of records for species listed in the United States Endangered Plant Species list and in the IUCN Red List of Threatened Plants. These two lists, however, are not comprehensive. If you are aware of other species that are better protected by limiting access to specific locality data, please contact us and we will update our database accordingly. We have also removed portions of information from records of species from groups such as Orchids, Cycads (including Zamiaceae) and Cacti,

> which are often subject to over-collection. Please note that we have removed specific locality information from all species in these groups, even those that are not explicitly listed as endangered.[3]

This "delicate balance" between complete unmediated online availability of collection information and responsible stewardship of potentially exploited organisms or objects is maintained in a variety of ways. Approaches range from suppression of entire records for specimens of federally listed endangered species, the approach taken by the Bivalve Collection at the Ohio State University of Biological Diversity, to the complete availability of all collection information for all specimens at the University of California, Berkeley Museum of Paleontology site.

The search screen for the Ohio mollusk site specifically states that "searches for federally endangered species will not return results" and interested users are advised that "just because no endangered species records are returned does not mean that they do not exist! If you require more specific locality information, including records for endangered species, contact the Curator."[4] While the University of California, Berkeley Museum of Paleontology site does provide complete online access to collection documentation, including accession, specimen, and locality records, it does caution that "these data should not be released prior to publication without approval by Museum staff."[5]

Other cautions on the use of the databased information fell into several broad categories. In addition to the intentionally incomplete representation of the collection as noted above, these included the inadequacy of the records as surrogates for the objects for research purposes, inconsistency of legacy data, and the lack of fixity of the information in the database.

Records as Surrogates. The online catalog records were not intended to serve as a substitute for actual examination of the specimen and its associated records. The American Museum of Natural History's Herpetology site stated that the curatorial staff was "alarmed at the practice of citing specimens and mapping localities from secondary sources, without verifying identifications."[6] Users were to view the online catalogs as tools to improve the efficiency of their onsite visits to the collections, enabling advance requests for specific material from the collection. One site saw the availability of specimen data as a form of publication intended to give "potential users of our collection advance information about material available by loan or for on-site study."[7] The American Museum of Natural History Paleontology site noted that "electronic collections are only proxies for the actual specimens" and the University of California, Berkeley Museum of Paleontology site urged investigators "to verify all data by direct examination of the specimens." However, in the case of virtual herbaria, such as that created by the Field Museum in collaboration with the Morton Arboretum, users have the benefit of a scanned image of the herbaria sheets themselves, complete with label, as well as the fielded data in the collection database. This allows the user the opportunity to make his own decisions on whether the information was properly transcribed or interpreted from the original source. Another online resource, the American Museum of Natural History's Web site for the Congo Expedition of 1909-1915, similarly includes both page images from the expedition's field notes, specimen records and published research to allow users to view the stages of research moving toward scientific publication.[8]

Inconsistency of Legacy Data. Inclusion of legacy data particularly that involving geographic names or other means of identifying specific collection locations, can add substantially to problems of inconsistent terminology. While internal users and collection managers may be aware of an individual collector's conventions of reporting data, external users may not. In the case of the collector who always used the location of his base camp for all locality data, regardless of how far afield the collections were actually made, not having this insight could significantly affect research results. It is difficult to capture both the data and an indication of its unreliability in the collection database. Retroactive interpretation of older collection records or legacy data, particularly when attempting to adhere to controlled vocabularies for subject terms and geographic names, requires considerable skill in the individual disciplines, knowledge of the information resource traditions in the specific museum, and may even require new taxonomic research.

Lack of Fixity of Database Contents. Users were reminded that the collection database is a work in progress, with ongoing additions and corrections as research continues on the existing specimen collections and new collections are added to the database. The American Museum of Natural History Paleontology site warned that "although the specimen database is regularly checked for discrepancies, there is no guarantee of its accuracy."[9] Several sites specifically requested that users notify the database administrators of errors and inconsistencies.

DISCUSSION

Historically, the museum collection management literature has encouraged museums to develop their own information systems and policies as appropriate to their individual collections and audiences. The development of commonly accepted descriptive standards, critical to the effective use of an unmediated online catalog, especially those information resources operating across multiple museum collections, was not emphasized prior to the automation of natural history collections.

Discussions of best practices of specimen description are seldom seen in the emerging body of museum collection management literature in the early 1900s, which deals largely with the management of office and other administrative files, rather than collection records. Introducing his 1927 guide to the management of small museums, Coleman specifically does not recommend standardization of practice, but intends to provide a "firm foundation for individuality."[10] Dudley and Wilkinson's 1958 *Museum Registration Methods*, arguably the first work intended to provide comprehensive coverage of museum collection management practices, was produced by registrars in the American Association of Museums. The contributors to this work, primarily registrars from art museums, had identified a need for consolidated information about registration procedures. The introduction states that the authors did not intend to "set up an ideal system of registration forms, but rather to give a representative selection of those currently in use."[11] The modern successor to this work, Buck and Gilmore's recent authoritative resource, *The New Museum Registration Methods*, does include chapters dealing with the processes and acquisitions, accessioning, and makes recommendations for policies governing access to collection records. This work cautions that access should be restricted to confidential material including donor and purchase information, but also insurance and appraisal histories, provenance, and condition and conservation treatment notes.[12] The 1994 Association for Systematics Collections guidelines for institutional policies recommends limited access to documentation dealing with "localities for economically valuable endangered species, fossils, and archaeological materials," particularly for non-academic users.[13] This work also specifically identifies "type localities, endangered species localities, insurance valuations [and] information concerning donors" as sensitive information requiring "protection from unauthorized users."[14]

Museums have a long tradition of non-uniform collection management practices. When discussing the development of museum collections and their associated documentation, Katherine and Philip Spiess note that "when the object or collection enters a museum, there is no single comprehensive approach guiding its management and use. Rather, each museum drawing from its own history and the traditions and cultures associated with its collection, and the experience of private and public collecting in its field, organizes and manages its collection to meet its own program needs."[15] A recent article on the rise of virtual museums remarked on the deep roots of these differences and suggested that museums extend this need to distinguish themselves to their users in both their physical as well as intellectual infrastructure, noting that "just as a physical museum is built with a certain architecture, museums have constructed different software frameworks to store the digital versions of their snakes and cockroaches."[16] However, unfavorable comparisons of museums to libraries focus on this lack of consistency. A recent Council on Library and Information Resources report identified the "heterogeneous recording practices that plague the museum community" as one of the greatest sustainability problems with resource-sharing initiatives that rely on museum collections.[17] In writing about the increasing demand for public accountability of museums, specifically access to museum information, Stam has acknowledged both the physical and intellectual scattering of museum records and the subsequent difficulties of access to "information stores in many locations, each based on a different set of principles, and with considerable overlap and lacunae."[18]

In a recent article arguing in support of electronic access to collection information as a means of information discovery, Sugden and Pennisi recognize the limitation of such isolated information, noting "data were held as card indexes and taxonomic monographs as well as physical specimens in museums and herbaria. Knowledge of specific taxonomic group was the privilege of handfuls of experts scattered in institutions on different continents."[19] Another *Science* editorial identified the "information technology challenge as liberating data from cabinets."[20]

One strong critic of the amalgamation of museum collection data on centralized multi-institutional sites cautions that this imposed uniformity obscures the original data and can lead to ungrounded statements about the range and distribution of plants and animals based on extrapolated data. McLaren notes that "whereas the field catalog or manual ledger may be diligent replications of the skin tag, database entries may be altered to conform to documentation standards. Fields such as specific locality may contain modified information for easier retrieval."[21] And this modification may result in incorrect conclusions. Another argument against the combination of individually created databases into a single resource is that errors of ambiguity may be introduced. Discussing the development of the Getty Art History Information Program, Bower wrote that "data that are unambiguous within the context of their initial capture . . . may become ambiguous when juxtaposed with data in different languages, data from different disciplines where overlapping terms have not been rendered referentially unique, or data from the same discipline that use different but equivalent terms to express names and concepts."[22]

In collaborative models such as HerpNET the participating museums contribute data to a centralized source, making their own decisions in advance as to what information they will supply. The HerpNET disclaimer particularly notes that the lack of data in the centralized catalog should not be interpreted as evidence that information about specific specimens or specific data about those specimens is lacking in the originating museum's collection catalog. Users of this data are urged to contact the holding museum for additional and supporting information, to confirm the existence of the specimen and to make their own determination of the accuracy of the data.

> The results returned by query forms on this site may contain fields that are not necessarily supported in the databases of every participating institution. The contents of these fields will appear blank in the records returned from these institutions. Similarly a value for a supported field may appear blank if no data were recorded in that field for a particular record. The extent to which data are made publicly available through this network is determined by each institution for its own data. In addition, every participating collection maintains additional specimen information that is not available through this interface. Individual institutions should be contacted for clarifications, and for additional data.[23]

This particular solution is similar to another collaborative catalog which provides access to potentially sensitive information, the Nazi-Era Provenance Internet Portal Project. This project has been designed by the American Association of Museums to provide improved access for both research and repatriation to art works in American museum collections. Nazi-Era works included are those that were created before 1946 and that were acquired after 1932, or that changed ownership between 1932 and 1946, or might be reasonably thought to have been in continental Europe between 1932 and 1946. While museums are urged to provide all 20 categories of object and provenance information listed in the project's provisions, museums may choose to limit disclosure on the Portal site and require that interested users contact the museum directly for further information.[24]

Another aspect of access to sensitive information that is not widely discussed in the literature is the response of museums to Freedom of Information (FOIA) requests. As many American museum collections include specimens collected on federal land, and therefore those collections are still owned by the federal governmental agencies even though they are housed in the nation's museums, it appeared likely that sensitive information could be requested under FOIA.[25] However, both anthropological collections and paleontological collections have specific exemptions from FOIA requests. Recent correspondence with a regional paleontologist with the Bureau of Land Management (BLM), which administers about 262 million acres of federal land which has been the source of many museum fossil collections, confirms that BLM "considers vertebrate fossil locality data as non-public information . . . Non-federal institutions which hold specimen collections from federal land and belonging to the federal government, would be subject to the same restrictions."[26] Again, access to sensitive information about federally owned collections is at the discretion of the individual museum curators.

CONCLUSIONS

Ownership of collections and their associated data can be manifested by limited physical and intellectual access to these materials. Natural history museums have long traditions of departmental autonomy, with museum functions and authority organized along discipline-based departmental lines. The evidence suggests that the traditional practices of access control have been translated into the online environment.

Based on the actual statements included on the collection database Web sites, only a few museums are overtly stating what they consider to be sensitive collection information. Fewer yet clearly state the criteria for its omission or suppression. Potential users of this condensed information are urged to examine the specimens and their associated documentation themselves and not accept without question the attributions and assumptions implicit in the catalog records. Users are also reminded that these catalogs are not static, i.e., not only that new material is being continuously added, but existing information is continuously being refined. At best, museums consider their online catalogs to be partial surrogates for the objects themselves.

Other as yet unstated reasons for withholding selected sensitive collection data may include prior agreements with the collector or donor restricting its availability to public users or the restriction of access to research data in active use by a museum staff member or visiting scholar.

Cataloging and recordkeeping systems originally developed for the internal use of museum staff as part of collection management may not translate easily to an online catalog to be used by unknown users. In addition to the challenge of creating records which are adequate surrogates for objects, creators and users of museum collection records must be aware of discipline-specific naming conventions, nomenclatural systems which change over time, incomplete catalog record information and debates over the public release of sensitive information about museum collections. Museums can no longer be working in isolation as the stewards of cultural heritage objects, but as cultural historian David Lowenthal reminds us, "issues of provenance, authenticity, protection, interpretation, display, commodification, legal title, restitution, repatriation, pillaging, illicit trade . . . are hardly ever tidily circumscribed within any single sphere; instead they require conjoined insights."[27]

NOTES

1. American Association of Museums. "The Accreditation Commission's Expectations Regarding Collections Stewardship." Approved 27 June 2001. Available at URL: http://www.aam-us.org/programs/accreditation/collstewardship.cfm [Accessed 29 December 2003].

2. These museums were the Academy of Natural Sciences of Philadelphia, American Museum of Natural History, Bell Museum of Natural History, California Academy of Sciences, Carnegie Museum of Natural History, Cleveland Museum of Natural History, Field Museum of Natural History, New York Botanical Garden, Missouri Botanical Garden, Museum of Comparative Zoology, National Museum of Natural History, Natural History Museum of Los Angeles County, Ohio State University Museum of Biological Diversity, Peabody Museum of Natural History, Provincial Museum of Alberta, Royal Ontario Museum, Royal Tyrrell Museum, University of California Berkeley Museum of Paleontology, University of Iowa Museum of Natural History, University of Kansas Natural History Museum and Biodiversity Center, University of Michigan Museum of Zoology, University of New Mexico Museum of Southwestern Biology, University of Texas at Austin Texas Memorial Museum.

3. New York Botanical Garden, "The Virtual Herbarium of the New York Botanical Garden." Available at URL http://sciweb.nybg.org/science2/endangered_species.html [Accessed 29 December 2003].

4. Ohio State University Museum of Biological Diversity, Molluscs Division, "OSA Freshwater Bivalve Collection." Last updated 2 June 2003. Available at URL: http://128.146.250.63/bivalves/ [Accessed 29 December 2003].

5. University of California, Berkeley Museum of Paleontology. "Collections Policies & Contacts." Available at URL: http://www.ucmp.berkeley.edu/collections/collpolicies.html#policies [Accessed 29 December 2003].

6. American Museum of Natural History, Department of Herpetology. "Collections." Available at URL: http://research.amnh.org.herpetology/collections/collections.html [Accessed 29 December 2003].

7. Field Museum, Invertebrates Collection Database. Available at URL: http://fml.fieldmuseum.org/collections/search.cgi?dest=inverts [Accessed 29 December 2003].

8. American Museum of Natural History, "Congo Expedition 1909-1915." Last updated January 2003. Available at URL: http://diglib1.amnh.org/ [Accessed 29 December 2003].

9. American Museum of Natural History, Division of Paleontology. Available at URL: http://paleo.amnh.org/fossil/how.html [Accessed 20 December 2003].

10. Laurence Vail Coleman, *Manual for Small Museums.* (New York: G.P. Putnam's Sons, 1927), 4.

11. Dorothy H. Dudley and Irma Bezold Wilkinson, eds. *Museum Registration Methods, 3rd ed. rev.* (Washington, DC: American Association of Museums, 1979).

12. Clarisse Carnell and Rebecca A. Buck, "Processes: Acquisitions and Accessioning," in *The New Museum Registration Methods*, ed. Rebecca A. Buck and Jean A. Gilmore (Washington, DC: American Association of Museum, 1998), 157-165.

13. K. Elaine Hoagland, ed., *Guidelines for Institutional Policies and Planning in Natural History Collections* (Washington, DC: Association of Systematics Collections, 1994), 22.

14. K. Elaine Hoagland, ed., *Guidelines for Institutional Policies and Planning in Natural History Collections* (Washington, DC: Association of Systematics Collections, 1994), 38.

15. Katherine and Philip Spiess, "Museum Collections," in *The Museum: A Reference Guide*, ed. Michael Shapiro (New York: Greenwood Press, 1990), 142.

16. James Gorman, "In Virtual Museums, An Archive of the World." *New York Times*, 12 January 2003. Available at URL: www.nytimes.com/2003/02/12/sciejce/12MUSE.html?pagewanted=print&position=top [Accessed 13 January 2003].

17. Diane Zorich, *A Survey of Digital Cultural Heritage Initiatives and their Sustainability Concerns* (Washington, DC: Council on Library and Information Resources, 2003), 25.

18. Deirdre Stam, "Public Access to Museum Information: Pressures and Policies," *Curator* 32 (1986): 192.

19. Andrew Sugden and Elizabeth Pennisi, "Diversity Digitized," *Science* 289 (2000): 2305.

20. Quentin D. Wheeler, Peter H. Raven and Edward O. Wilson, "Taxonomy: Impediment or Expedient?" *Science* 303 (2004): 285.

21. Suzanne B. McLaren, "On Database Accessibility."*ASC Newsletter* [Association of Systematics Collections] 20 (December 1992): 159.

22. James M. Bower, "Vocabulary Control and the Virtual Database," *Knowledge Organization* 20 (1993): 5.

23. HerpNET. Available at URL: http://herpnet.digir.net/pres/PresentationServlet?action=home [Accessed 11 December 2003].

24. Nazi-Era Provenance Internet Portal Project. "Who Should Participate?" Available at URL: http://www.nepip.org/public/museuminfo/participate.cfm?menu_type=museuminfo#foQ4 [Accessed 29 December 2003].

25. FOIA Reference Guide, Revised November 2003. Available at URL: http://www.usdoj.gov/04foia/04_3.html [Accessed 20 September 2004].

26. Personal communication with Laurie Bryant, 1 December 2003.

27. David Lowenthal, "Pioneering Stewardship: New Challenges for CRM," *CRM: The Journal of Heritage Stewardship* 1 (Fall 2003): 8-9.

Index

Page numbers followed by n indicate notes.

BOOK ORDER FORM!

Order a copy of this book with this form or online at:
http://www.haworthpress.com/store/product.asp?sku=5617

Collaborative Access to Virtual Museum Collection Information

Seeing Through the Walls

____ in softbound at $17.95 ISBN-13: 978-0-7890-2934-8. / ISBN-10: 0-7890-2934-0.
____ in hardbound at $34.95 ISBN-13: 978-0-7890-2933-1. / ISBN-10: 0-7890-2933-2.

COST OF BOOKS ________

POSTAGE & HANDLING ________

US: $4.00 for first book & $1.50 for each additional book
Outside US: $5.00 for first book & $2.00 for each additional book.

SUBTOTAL ________

In Canada: add 7% GST. ________

STATE TAX ________

CA, IL, IN, MN, NJ, NY, OH, PA & SD residents please add appropriate local sales tax.

FINAL TOTAL ________

If paying in Canadian funds, convert using the current exchange rate, UNESCO coupons welcome.

❑ **BILL ME LATER:**

Bill-me option is good on US/Canada/ Mexico orders only; not good to jobbers, wholesalers, or subscription agencies.

❑ **Signature** ________

❑ **Payment Enclosed: $** ________

❑ **PLEASE CHARGE TO MY CREDIT CARD:**

❑ Visa ❑ MasterCard ❑ AmEx ❑ Discover
❑ Diner's Club ❑ Eurocard ❑ JCB

Account # ________

Exp Date ________

Signature ________

(Prices in US dollars and subject to change without notice.)

PLEASE PRINT ALL INFORMATION OR ATTACH YOUR BUSINESS CARD

Name

Address

City State/Province Zip/Postal Code

Country

Tel Fax

E-Mail

May we use your e-mail address for confirmations and other types of information? ❑ Yes ❑ No We appreciate receiving your e-mail address. Haworth would like to e-mail special discount offers to you, as a preferred customer.
We will never share, rent, or exchange your e-mail address. We regard such actions as an invasion of your privacy.

Order from your **local bookstore** or directly from
The Haworth Press, Inc. 10 Alice Street, Binghamton, New York 13904-1580 • USA
Call our toll-free number (1-800-429-6784) / Outside US/Canada: (607) 722-5857
Fax: 1-800-895-0582 / Outside US/Canada: (607) 771-0012
E-mail your order to us: orders@haworthpress.com

For orders outside US and Canada, you may wish to order through your local sales representative, distributor, or bookseller.
For information, see http://haworthpress.com/distributors

(Discounts are available for individual orders in US and Canada only, not booksellers/distributors.)

Please photocopy this form for your personal use.
www.HaworthPress.com

BOF05